REFUSING ECOCIDE

Refusing Ecocide: From Fossil Capitalism to a Liveable World provides a critical analysis of the central role of fossil capitalism in causing climate change and argues that only alternatives based upon democratic eco-socialism can prevent the deepening of the climate crisis.

Employing three core concepts within historical materialism – capitalist accumulation, imperialism and hegemony – it locates the existential threat of our changing climate in the drive for increasing profit and growth, the domination of advanced capitalist states that strip resources and exploit cheap labour, and the consent to the capitalist way of life in the global North. With attention to the ways in which, powered by fossil fuels, capital has subjected the world to its predatory logic, this book charts this history and surveys the damage from the Industrial Revolution to today's deep civilizational crisis, arguing that the market-based and purely technological solutions of 'climate capitalism' are too little, too late.

A call for a multifaceted and multi-scalar shift away from capitalist accumulation, imperialism and class hegemony and instead towards democratic eco-socialism, it will appeal to scholars across the social sciences with interests in political and social theory, the environment and sustainability.

William K. Carroll is Professor of Sociology at the University of Victoria, Canada. His research explores relationships between corporate power, fossil capitalism and the climate crisis, the political economy and ecology of corporate capitalism, social movements and social change, and critical social theory and method. He has also co-directed 'Mapping the power of the carbon-extractive corporate resource sector', a partnership of several universities and civil-society organizations which has examined corporate power and resistance within the global political economy with a focus on fossil capital based in western Canada. He is the author of *Expose, Oppose, Propose: Alternative Policy Groups and the Struggle for Global Justice* (2016) and *The Making of a Transnational Capitalist Class: Corporate Power in the 21st Century* (2010) and the co-author of *Organizing the 1%: How Corporate Power Works* (2018). He is also the editor of *The Elgar Companion to Antonio Gramsci* (2024) and *Regime of Obstruction: How Corporate Power Blocks Energy Democracy* (2021) and the co-editor of *A World to Win: Contemporary Social Movements and Counter-Hegemony* (2016).

Critiques and Alternatives to Capitalism

Series editor: Marcello Musto

Professor of Sociological Theory, York University, Canada

This series publishes scholarly works on critiques and alternatives to capitalism, spanning a number of subject matters, political perspectives and geographical areas. It welcomes monographs and edited volumes in the fields of sociology, social and political theory and heterodox political economy, whose main areas of focus are the problems of capitalist society and its mode of production; alternatives to capitalism that address contemporary social issues; and 19th and 20th centuries anti-capitalist ideas and practical experiments.

Welcoming new perspectives on a wide range of themes, it seeks to explore alternative social-economic systems, critical theories of capitalism, social classes and inequality, public/private ownership and new contours of 'the commons', economic and financial crises, ecology, globalization, migration and citizenship, gender oppression, alienation and cultural critique. The result is an eclectic, but focussed and informative, series that provides original investigations, inspires significant conversations for today and appeals to a diverse international audience.

Life Imaginaries, Environmental Rationality, and Dialogue of *Savoirs*
The Sustainability of Life
Enrique Leff

Ten Essays for a 21st Century Socialism
Tony Andréani

Refusing Ecocide
From Fossil Capitalism to a Liveable World
William K. Carroll

For more information about this series, please visit: www.routledge.com/Critiques-and-Alternatives-to-Capitalism/book-series/CAATC

REFUSING ECOCIDE

From Fossil Capitalism to a Liveable World

William K. Carroll

Routledge
Taylor & Francis Group
LONDON AND NEW YORK

Designed cover image: Getty Images

First published 2025
by Routledge
4 Park Square, Milton Park, Abingdon, Oxon OX14 4RN

and by Routledge
605 Third Avenue, New York, NY 10158

Routledge is an imprint of the Taylor & Francis Group, an informa business

British Library Cataloguing-in-Publication Data
A catalogue record for this book is available from the British Library

ISBN: 978-1-032-53641-5 (hbk)
ISBN: 978-1-032-53642-2 (pbk)
ISBN: 978-1-003-41292-2 (ebk)

DOI: 10.4324/9781003412922

Typeset in Galliard
by SPi Technologies India Pvt Ltd (Straive)

For my grandsons, Kaito and Tomohisa,
and for all our grandchildren.

CONTENTS

ILLUSTRATIONS

Figures

Tables

ACKNOWLEDGEMENTS

Although it is a single-authored work, this book is very much a collective product. It stems from my engagement with a superb team of 100-odd scholars and activists who formed the Corporate Mapping Project, a research and community engagement project, supported by the Social Sciences and Humanities Research Council of Canada, from 2015 through 2023. The Project, which I co-directed with Shannon Daub, combined social science and social action in mapping relations of power and resistance in and around Canada's outsized fossil-fuel sector. I am grateful to the entire team, from whom I have learned so very much. I appreciate the invitation from Marcello Musto, Editor of Routledge's Critiques and Alternatives to Capitalism, to contribute to the series, and the support of key editorial staff at Routledge in what has been a very smooth production process – in particular, Neil Jordan, Alice Salt, Gemma Rogers, Chris Parry and Kriti Rana. Finn Deschner and Babak Ashrafkhani Limoudehi provided indispensable research assistance throughout the fall and winter of 2023/2024, and Finn continued through the spring of 2024 as an editorial assistant in preparing the final manuscript. Finally, I thank several colleagues who have given permission for me to reprint graphics from their publications: Andreas Malm (for Figure 1.1), Geoff Mann (for Table 5.1) and David Ciplet (for Figure 6.1).

Victoria, 7 June 2024

INTRODUCTION

As I write this introduction in May 2024, the habitability of our planet for a human population of eight billion is in peril.

The symptoms of climate breakdown are undeniable.

On February 8, 2024, Earth reached a new milestone. The European Union's Copernicus Climate Change Service reported that mean temperature in the previous 12 months had been more than 1.5 degrees Celsius above pre-industrial levels. As the Progressive International (2024) commented, 'just nine short years ago, the world's governments agreed in Paris that they would limit global warming to below 1.5 degrees. "1.5 to stay alive" was the mantra. They failed in record time.' In fact, the World Meteorological Organization's *State of Global Climate 2023*, issued in March 2024, reports a plethora of records broken, even smashed, for greenhouse gas levels, ocean heat and acidification, surface temperatures, sea level rise, glacier retreat and Antarctic sea ice cover (World Meteorological Organization 2024). These 'records' are broken at our peril, and indeed, the human dimensions of climate change are also undeniable. As the report observes, in 2023 'heatwaves, floods, droughts, wildfires and intense tropical cyclones wreaked havoc on every continent and caused huge socioeconomic losses' (World Meteorological Organization 2024:iii). The consequences were particularly devastating for vulnerable populations, as 'extreme climate conditions exacerbated humanitarian crises, with millions experiencing acute food insecurity and hundreds of thousands displaced from their homes' (World Meteorological Organization 2024). It is hardly surprising that nearly 80 percent of the world's leading climate scientists, when polled in the spring of 2024, foresaw a global temperature increase of at least 2.5 degrees Celsius by end of century. The vast

DOI: 10.4324/9781003412922-1

majority of the most knowledgeable people on our planet 'expect climate havoc to unfold in the coming decades' (Carrington 2024).

In 2018, a team of earth scientists introduced the term 'Hothouse Earth', in a widely cited article. They noted a rapid advance towards planetary thresholds at which 'intrinsic biogeophysical feedbacks in the Earth System ... could become the dominant processes controlling the system's trajectory' (Steffen et al. 2018:8254). These thresholds are called 'tipping points', beyond which climate change accelerates and becomes irreversible, in a cascade of positive feedback. The feedback processes include permafrost thawing, which releases methane (a greenhouse gas 84 times more impactful than CO_2) and loss of polar ice sheets, which is not only elevating sea levels but weakening the albedo effect.[1] As the planet warms, increased atmospheric methane and shrinking polar ice caps, triggered by global heating, amplify global heating.

There are other feedback loops. Climate chaos already underway (drought, wildfires) contributes to the die-back of tropical and boreal forests – turning ecosystems that have fixed carbon for millennia into grasslands and carbon bombs. Meanwhile, increasing volumes of atmospheric carbon precipitate as acid rain, reducing the oceans' capacity to absorb carbon, and further heating the hothouse.[2] Steffen et al noted that 'Hothouse Earth is likely to be uncontrollable and dangerous to many ... and it poses severe risks for health, economies, political stability ... and ultimately, the habitability of the planet for humans' (2018:8257). The challenge for humanity is to create 'a "Stabilized Earth" pathway that steers the Earth System away from its current trajectory toward the threshold beyond which is Hothouse Earth' (2018:8254).

The situation has been described as climate crisis, climate breakdown, climate emergency, but, most ominously, as ecocide. In the scholarly literature, 'ecocide' emerged as a term in 1970 (Weisberg 1970), but was rarely invoked in the titles of academic articles and books until recent years.[3] In his eponymously titled book, David Whyte defined ecocide as 'the deliberate destruction of our natural environment' (2020:2), and pointed his finger directly at the profit-driven corporations that 'are wrecking our world' (2020). Stop Ecocide International, a movement organization campaigning to make ecocide an international crime, defines the term as 'the mass damage and destruction of the natural living world,' literally '"killing one's home"' (Stop Ecocide n.d.). Ecocide does not mean the end of nature, which is indifferent to any particular species, including ours. Nor is the prospect of ecocide a death sentence for all of humanity. Just as genocide – a term painfully familiar to us from contemporary Palestine – does not mean the annihilation of an entire ethnic group, ecocide implies the destruction of the conditions for a decent life for vast numbers of people. The growing numbers of climate refugees from regions that are already becoming uninhabitable – whether from sea level rise or desertification – confirm that ecocide is already upon us, in its earliest stage. Ecocide is very much about the Earth's prospects to support a large population of humans, but other species are already in sharp decline. Drawing on the most recent figures from the Living Planet Index,[4] Patrick Greenfield (2022) notes that between 1970 and 2018 'wildlife populations have plunged by an average of 69%'; 'the abundance of mammals, birds, fish, amphibians and reptiles is falling fast, as populations of sea lions, sharks, frogs and salmon collapse.' Indeed, the climate crisis is entirely entangled with the Sixth Extinction, the cumulative decline of living systems – and of biodiversity – as forests are cleared to make room for cattle grazing, reducing biodiversity, impairing the Earth's 'lungs' from fixing atmospheric carbon and increasing carbon emissions. Elizabeth Kolbert writes, in her Pulitzer Prize winning *The Sixth Extinction*,

> Having freed ourselves from the constraints of evolution, humans nevertheless remain dependent on the earth's biological and geochemical systems. By disrupting these systems – cutting down tropical forests, altering the composition of the atmosphere, acidifying the oceans – we're putting our own survival in danger.
>
> *(2014:267)*

Although the *Sixth Extinction* looms in the background, this book is laser-focussed on the single most urgent ecological and existential challenge of our time: the climate crisis, its human causes and the possibilities for refusing this cardinal element of ecocide.

The climate crisis is a natural phenomenon, driven by human activities (which, as I will explain in Chapter 1, are also natural phenomena). This has led many earth scientists to argue that the planet has crossed into a new epoch, the Anthropocene – that of the 'geology of mankind' – (Crutzen 2002), 'beginning around 1950, representing the emergence of human-industrialized society as the primary factor in Earth System change' (Foster 2024:249). Critical social scientists, however, have noted that this term can be misleading. It fails to identify the specific human agents, operating within social structures, who are the primary drivers of the crisis. Andreas Malm (2016:391) suggests, 'this is the geology not of mankind, but of capital accumulation.'

Certainly, the corporations that dominate the fossil fuel sector, and their predecessor firms, are the major culprits. In 2014, Richard Heede published an article documenting that, between 1751 (before the invention of the steam engine) and 2010 the 90 biggest emitters – the 'carbon majors' – were responsible for 63% of cumulative worldwide emissions of greenhouse gases, with half of all emissions having been released since 1986. In the seven years following the 2015 Paris climate accord, the carbon majors increased their emissions; indeed 80% of global emissions from 2016 through 2022 'can be traced to just 57 corporate and state producing entities', demonstrating 'the outsized influence of a small group of producers that are increasing production' (InfluenceMap 2024:4, 26). Their investment decisions have directly caused the climate crisis, and continue to exacerbate it. The CEOs of these fossil-fuel producers could fit into a couple of school buses.

But the carbon majors are enabled by other actors and institutions. At the Corporate Mapping Project, a community-university research and public engagement project I co-directed from 2015 to 2023 (Carroll and Daub 2015; Carroll 2021), we mapped the various connections between the fossil-fuel sector and the economic, political and cultural organizations that enable and legitimate its activities, focussing on the case of Canada, a major fossil-fuel producer. In Chapter 4, I reflect on some of our findings, which reveal a 'regime of obstruction' that enables and protects the revenue streams and fixed-capital investments of fossil-fuel corporations and blocks meaningful climate action. The key actors in this regime include the banks and institutional investors that finance the industry, the captured regulators that green-light new investments, the government ministries in regular dialogue with industry lobbyists and a range of organizations that legitimate the industry, from think tanks, industry groups and business councils to corporate and astro-turf media and business schools.

This book focusses on these centres of economic, political and cultural power. I analyze how their actions have driven the climate crisis, and why the ruling economic and political bloc, based in the advanced capitalist west, or global North, is incapable of steering humanity to a safe destination. Ecocide, however, is not inevitable. In recent decades many initiatives have appeared in resistance and opposition to ecocidal practices. These include:

- grassroots movements and street actions to raise consciousness and pressure governing elites, such as Extinction Rebellion and Fridays for Future;
- more organized popular opposition grounded in the left, such as like Trade Unions for Energy Democracy, the Global Ecosocialist Network and the Climate Justice Alliance;
- online platforms that offer critical policy analysis focussed on the fossil-fuel industry and its allies, enablers and legitimators, such as Oil Change International, InfluenceMap and the Climate Social Science Network; and
- alternative media focussed on the ecological crisis, such as Desmog, Green Left and Climate&Capitalism.

However, in the dominant institutions, including the annual Convention of the Parties (COP) that governs the United Nations Framework Convention on Climate Change, corporate interests predominate, fettering the transformations that are obviously necessary. At COP 28, held in Dubai in December 2023, a record number of fossil fuel lobbyists (2,456 in all) outnumbered all national delegations but those of the host and host-designate. Dwarfing all official Indigenous representatives 7-to-1, those lobbyists found ready allies among the many fossil-fuel employees embedded in the national delegations of France, Italy and other Northern countries (Corporate Europe Observatory 2023). Not surprisingly, COP 28, like all previous COPs,

> failed to agree on the need to phase out fossil fuels and to set a deadline for doing so. Instead, the final document suggests that states may – with no obligations – 'draw down' fossil fuel production. The demands from over 120 countries to completely eliminate new fossil fuel production were ignored.
>
> *(Progressive International 2023)*

The chasm between the urgency of serious climate action, understood by many, and the stasis of climate policy at global and national levels is palpable. Why this stasis, and what is to be done?

This book addresses these questions, in two parts. The first develops a critical perspective on fossil capitalism and climate breakdown, drawing primarily on the historical materialist framework first introduced by Karl Marx and Friedrich Engels. This framework directs our attention towards a 'trifecta of power' at the centre of capitalism as a way of life, illuminated via three core concepts. Capitalist *accumulation*, the economic aspect, is driven by the private profit motive towards endless growth and increasing inequities. *Imperialism*, the geo-political-economic aspect, entails the domination of the global South by advanced capitalist states as they strip the land of resources and super-exploit cheap labour. Finally, *hegemony*, the cultural and political aspect, organizes popular consent to the capitalist way of life, particularly in the global North.

In Chapter 1, I unpack that trifecta, and apply it in recounting the century-and-a-half epoch from the Industrial Revolution to the close of World War Two, during which fossil capitalism was fully established. Chapter 2 takes up the post-war era – the 'golden years' of economic prosperity, consumer capitalism and class compromise, from the mid-1940s into the 1970s, during which the Great Acceleration took shape, exponentially increasing carbon emissions and trending towards ecological overshoot.[5] The final chapter in Part 1 brings us up to date. Powered by fossil fuels, capital has subjected the world to its predatory logic, eventuating in today's deep civilizational crisis. In the 1980s and 1990s, the economic crisis of the 1970s was resolved through neoliberal policies of deregulation, privatization and austerity, as capital became more globalized and financialized. Although the global financial meltdown of 2008 discredited neoliberalism, no alternative has gained favour in the centres political-economic power. As climate breakdown becomes increasingly visible, and as a diverse array of movements for climate justice and sanity intensify hegemonic struggle, the neoliberal zombie stumbles towards the precipice.

The book's second part appraises solutions to the climate crisis that have been proposed from various quarters. In these chapters I examine how proposed remedies square with our core analytical concepts of accumulation, imperialism and hegemony. This engagement with alternatives begins in Chapter 4 with a critique of the 'false solutions' comprising Climate Capitalism: attempts to regulate fossil capital via market mechanisms or to create techno-fixes, as in changing the energy source without addressing the socio-ecological relations at the heart of the crisis. The rejection of these approaches, which remain within the logic of accumulation, imperialism and bourgeois hegemony, leads to the final two chapters. Chapter 5 explores three alternatives that move us in the right direction, yet fail to provide the comprehensive approach that the civilizational and climate crisis actually demands. These projects – the Green New Deal, Degrowth, and Buen Vivir – each contain currents that could converge upon a refusal of ecocide. The challenge lies in pulling these social forces into a coherent hegemonic project with a mass base. Chapter 6 takes up this challenge. I argue for a democratic eco-socialism that breaks decisively from both fossil capitalism and Climate Capitalism. A capacious eco-socialist project directly confronts the trifecta of power that is at the heart of ecological degradation and social injustice. It is our best bet, against lengthening odds, in refusing ecocide.

Notes

1 As Kashiwase et al. (2017:8170) explain, 'Ice-albedo feedback is a key aspect of global climate change. In the polar region, a decrease of snow and ice area results in a decrease of surface albedo, and the intensified solar heating further decreases the snow and ice area.'
2 'Each year, about one third of the carbon dioxide (CO_2) in fossil fuel emissions dissolves in ocean surface waters, forming carbonic acid and increasing ocean acidity. Over the next century or so, acidification will be intensified near the surface where much of the marine life that humans depend upon live' (Canada 2011:10).
3 From 1970 to 1990 'ecocide' appeared in titles of 49 articles and books in the Google Scholar database. The term was used in titles of 187 publications between 1991 and 2010. From 2011 to 25 April 2024, 770 academic publications included 'ecocide' in their titles (393 since 2021; search conducted on 25 April 2024).
4 https://www.livingplanetindex.org/ accessed 26 April 2024.
5 The Global Footprint Network estimates that 'humanity is using nature 1.7 times faster than our planet's biocapacity can regenerate. That's equivalent to using the resources of 1.7 Earths.' For everyone on the planet to live like the average American or Canadian would require 5.1 Earths (Earth Overshoot Day 2024).

References

Canada. 2011. *Canada's State of the Oceans Report, 2012*. Ottawa: Fisheries and Oceans Canada.

Carrington, Damian. 2024. "World's Top Climate Scientists Expect Global Heating to Blast Past 1.5C Target." *The Guardian*. Retrieved May 13, 2024 (https://www.theguardian.com/environment/article/2024/may/08/world-scientists-climate-failure-survey-global-temperature).

Carroll, William K. 2021. "Introduction." in *Regime of Obstruction*, edited by W. K. Carroll. Edmonton: AU Press, pp. 3–31.

Carroll, William K., and Shannon Daub. 2015. "We're Putting Fossil Fuel Industry Influence under the Microscope." *Corporate Mapping Project*. Retrieved May 1, 2024 (https://www.corporatemapping.ca/were-putting-fossil-fuel-industry-influence-under-the-microscope/).

Corporate Europe Observatory. 2023. "Record Number of Fossil Fuel Lobbyists Granted Access to COP28 Climate Talks." *Corporate Europe Observatory*. Retrieved May 2, 2024 (https://www.corporateeurope.org/en/2023/12/record-number-fossil-fuel-lobbyists-granted-access-cop28-climate-talks-0).

Crutzen, Paul J. 2002. "Geology of Mankind." *Nature* 415(6867):23–23. doi: 10.1038/415023a

Earth Overshoot Day. 2024. "How Many Earths? How Many Countries?" *Earth Overshoot Day*. Retrieved April 30, 2024 (https://overshoot.footprintnetwork.org/how-many-earths-or-countries-do-we-need/).

Foster, John Bellamy. 2024. *The Dialectics of Ecology: Socialism and Nature*. New York: Monthly Review Press.

Greenfield, Patrick. 2022. "The Biodiversity Crisis in Numbers – a Visual Guide." *The Guardian*. Retrieved April 26, 2024 (https://www.theguardian.com/environment/2022/dec/06/the-biodiversity-crisis-in-numbers-a-visual-guide-aoe).

Heede, Richard. 2014. "Tracing Anthropogenic Carbon Dioxide and Methane Emissions to Fossil Fuel and Cement Producers, 1854–2010." *Climatic Change* 122(1–2):229–41. doi: 10.1007/s10584-013-0986-y

InfluenceMap. 2024. "The Carbon Majors Database: Launch Report." Retrieved May 28, 2024 (https://carbonmajors.org/briefing/The-Carbon-Majors-Database-26913).

Kashiwase, Haruhiko, Kay I. Ohshima, Sohey Nihashi, and Hajo Eicken. 2017. "Evidence for Ice-Ocean Albedo Feedback in the Arctic Ocean Shifting to a Seasonal Ice Zone." *Scientific Reports* 7(1):8170. doi: 10.1038/s41598-017-08467-z

Kolbert, Elizabeth. 2014. *The Sixth Extinction*. New York: Henry Holt and Company.

Malm, Andreas. 2016. *Fossil Capital: The Rise of Steam Power and the Roots of Global Warming*. London: Verso.

Progressive International. 2023. "PI Briefing No. 50 COP Out." *Progressive International*. Retrieved May 2, 2024 (https://progressive.international/wire/2023-12-15-pi-briefing-no-50-cop-out/en).

Progressive International. 2024. "PI Briefing | No. 6 | 1.5 to Stay Alive?" *Progressive International*. Retrieved May 28, 2024 (https://progressive.international/wire/2024-02-09-pi-briefing-no-6-1-5-to-stay-alive/en).

Steffen, Will, Johan Rockström, Katherine Richardson, Timothy M. Lenton, Carl Folke, Diana Liverman, Colin P. Summerhayes, Anthony D. Barnosky, Sarah E. Cornell, Michel Crucifix, Jonathan F. Donges, Ingo Fetzer, Steven J. Lade, Marten Scheffer, Ricarda Winkelmann, and Hans Joachim Schellnhuber. 2018. "Trajectories of the Earth System in the Anthropocene." *Proceedings of the National Academy of Sciences* 115(33):8252–59. doi: 10.1073/pnas.1810141115

Stop Ecocide. n.d. "What Is Ecocide?" *Stop Ecocide*. Retrieved May 28, 2024 (https://www.stopecocide.earth/what-is-ecocide).
Weisberg, Barry. 1970. *Ecocide in Indochina: The Ecology of War*. San Francisco: Canfield Press.
Whyte, David. 2020. *Ecocide: Kill the Corporation before It Kills Us*. Manchester: Manchester University Press.
World Meteorological Organization. 2024. "State of the Global Climate 2023." Retrieved May 2, 2024 (https://library.wmo.int/records/item/68835-state-of-the-global-climate-2023).

PART I

Fossil Capitalism and Climate Crisis

1

FOSSIL CAPITALISM AND THE TRIFECTA OF POWER

The findings of climate science are by now unassailable, and future scenarios, projecting from those findings, are grim – even within a narrow window of, say, a decade or so. The extreme weather we have been experiencing in various forms, a basic symptom of climate breakdown, confirms our entrance into an era of cascading ecological crises. The issue is whether humanity can salvage from the wreckage the basis for a liveable world.

Climate science, a field of natural science that includes oceanography, atmospheric science and glaciology, cannot on its own address this issue. If the climate crisis has been generated by human activity, effective responses to heal the earth system must be informed both by climate science and by social science. Obviously, the practices driving ecological degradation occur within specific, historically formed social relations. As the crisis has deepened, social scientists have turned their attention to the political, economic and psycho-cultural forces that have shaped our ecocidal way of life.

The voluminous literature that has emerged since the 1970s is far too extensive to survey here, and that is not my objective. My presentation will be selective, and synthetic, beginning with a framework introduced by Allan Schnaiberg in 1980 that has been a touchstone for critical political ecology, namely the treadmill of production. Schnaiberg's treadmill clearly delineates the ecological implications of the endless growth that capital requires, and directs us towards one important strand of thought that offers a particularly powerful perspective, capable of integrating insights from both natural science and social science. That perspective is historical materialism (HM).

DOI: 10.4324/9781003412922-3

A Trifecta of Power: Accumulation, Imperialism and Hegemony

In this chapter I unfold three key theoretical concepts that have been central to HM: *accumulation, imperialism and hegemony*. Together, they enable us to discern how power operates within contemporary capitalism. They thereby provide a framework for subsequent chapters, which track the development of Fossil Capitalism and the climate crisis and assess a range of proposed solutions.

Let us begin the discussion by taking up capital accumulation and its problematic relationship to the natural world. Schnaiberg's treadmill of production provides a good entry-point. Like biologist Rachel Carson's (2002 [1962]) *Silent Spring*, often noted as an impetus for modern environmentalism, Schnaiberg's text drew upon natural science in linking industrialism to the degradation of ecosystems. Building on Carson, Schnaiberg, a social scientist, went further. Lamenting the 'woeful ignorance…of the systematic connections between our social and economic structures, and the productive organization of the physical, biotic environment,' Schnaiberg presented a 'social structural view of the environment' (1980:12, 9), tracing a clear linkage between the corporate pursuit of profit and environmental degradation. He observed that competition for profit and market share compels industrial corporations to expand production, and to re-invest profits in high technology, increasing labour productivity but decreasing labour input to production, as technology replaces living labour. As corporations grow in this way, the expanding volume of produced commodities must be absorbed by consumers – by developing markets abroad and also through higher wages paid to workers (primarily in the global North). Thus, 'competition and the quest for profitability constitute the main construction materials for the production treadmill' (Schnaiberg 1980:230).

As corporations re-invest their profits in expanded production, replacing living labour with labour-saving technologies, not only does the scale of the industrial economy grow, but the treadmill of production must accelerate to keep the same number of workers (who are also consumers) employed – let alone grow the employed workforce. For Schnaiberg, the ecological implications were clear. As he succinctly put matters, '*increasing the speed of this treadmill involves increased environmental withdrawals and additions*' (1980:230). *Withdrawals* refer to extraction of natural resources, which deplete ecosystems by removing integral aspects of living systems – for instance, 'mountaintop removal' as an industrial mining practice. *Additions* refer to the various forms of pollution that are by-products of industrial production and that further degrade ecosystems – as in the dumping of plastics in the ocean. In short, 'the accumulation of capital relies on nature – both as a "tap," which supplies material and energic inputs to commodity production, and as a "sink" for absorbing the latter's waste' (Fraser 2022:118). As the treadmill accelerates, so too does the ecological crisis.

In their appraisal of the treadmill framework a quarter of a century after the initial publication, Schnaiberg and his colleagues pointed out that *production* has analytical priority because 'it is in the decision to provide supply, and the means by which that supply is provided, where social systems and ecosystems first collide' (Kenneth et al. 2004). This directs our attention to the people and organizations that wield power in enabling and controlling investment and in managing production – the so-called 'treadmill elites' (ibid.: 302). Meanwhile, in the global North, replacement of workers with technology-intensive production not only led to an acceleration of the treadmill; it generated increased social support for further acceleration, to protect existing jobs (ibid.: 297).

Although consumption is subordinate to production (only that which has been produced can be consumed), production and consumption practices are mutually reinforcing. Within the culture of consumer capitalism that began to flourish during the three 'golden decades' following World War Two (see Chapter 2), consumer goods, including automobiles, appliances and houses, are identified with success, affluence and comfort. Workers, particularly if they are atomized – lacking the solidaristic ties that issue from union membership and the like – come to see themselves as competing with each other for jobs, income and the markers of success (Lebowitz 2020). They don golden straitjackets, and may view environmental protection as an existential threat, as in the binary of 'jobs versus environment'.

From the Treadmill of Production to the Metabolic Rift

In identifying the ecological implications of capitalism's in-built growth imperative the treadmill of production provides an entry-point for our analysis of capitalism and ecological crisis. But as Clark and York (2005:18) point out, the theory 'says little about the system of capitalism as it pursues endless accumulation of capital and divides nature and humanity for the sake of profit.' That system has been the object of a transdisciplinary perspective known as historical materialism, or simply, Marxism.

The originators of historical materialism were Karl Marx (1818–1883) and Friedrich Engels (1819–1895). Their collaborative scholarship in works such as *The German Ideology* (1965 [1846]) and, most famously, the *Communist Manifesto* (2019 [1848]), launched a wide-ranging programme of research and theorizing in the service of social critique and political action.

From the start, historical materialism differed from other approaches within social science, on two counts.

On the one hand, as a theoretical and philosophical perspective, HM has emphasized the emergent, dynamic, relational and material character of the social world, and its embeddedness within the larger natural world. In this perspective, 'nature and society are not different realities, but are co-evolving existences, in which society is asymmetrically dependent upon the larger natural world of which it is a part' (Foster 2022). As Marx reflected in 1844,

> Nature is man's *inorganic* body – nature, that is, insofar as it is not itself human body. Man *lives* on nature – means that nature is his body, with which he must remain in continuous interchange if he is not to die. That man's physical and spiritual life is linked to nature means simply that nature is linked to itself, for man is a part of nature.
>
> *(Marx 1959 [1844]:31)*

To comprehend the human condition we must conceptually grasp the relations among people and between humanity and nonhuman nature as they have emerged in natural history (which itself includes human history), and as they are reproduced today – attending in our analysis to the contradictions in those relations which give impetus towards change.

In capitalist modernity, these relations comprise a system of 'generalized commodity production' (Mandel 1976:20): not only do goods and services take a commodity form, but the human agency that produces those goods and services is bought and sold on labour markets, as a commodity. Wage labour is both premise and result of capitalism, a system of commodity production and consumption that is incessantly reproduced as capital employs labour and as labour produces capital (both the consumer goods that provide subsistence and the means of production that provide the basis for further production). In this conception, as Michael Lebowitz recounts, 'capitalism is an organic system because it reproduces its premises—the capitalist and the wage-laborer...' (2020:44).

On the other hand, HM has insisted on the need to develop critical knowledge that can inform critical action. As Marx declared in 1843, at the age of 25, 'there can still be no doubt about the task confronting us at present: the *ruthless criticism of the existing order*, ruthless in that it will shrink neither from its own discoveries, nor from conflict with the powers that be' (Marx 1843). The close link HM draws between consciousness and action is most concisely expressed in Marx's eleventh Thesis on Feuerbach – that 'philosophers have hitherto only *interpreted* the world, in various ways; the point is to *change* it' (Marx 2002 [1845]). This commitment to *praxis* – to developing an understanding of our world as we participate in its transformation (which also entails transforming ourselves) has meant that, for the most part, HM has been influential as a framework informing progressive movements, yet confined to the margins of dominant institutions within capitalist society.

Marx and Engels were not academics, and neither were most Marxist intellectuals, well into the 20th century. As a critical theory directly tied to emancipatory practice, HM found its institutional basis in socialist and communist political parties and movements. In the eyes of liberal and conservative intellectuals, this in itself discredited HM as science.[1] And, it must be admitted, one stream of Marxism departed from the critical, questioning practices of dialectical analysis, positing instead a bright socialist future following the inevitable collapse of capitalism (Larrain 1986). Although leftist movements, and Marxism itself, were actively repressed during the first Cold War in the 1950s, the emergence of New Left movements in the 1960s led to a revival of HM and opened space within academe for Marxist and other radical perspectives. That space has widened as the contradictions of globalizing capitalism have become more salient, inspiring new waves of activism and critical scholarship (McNally 2006).

The historical materialism I draw upon here posits no teleology of history. Instead, it emphasizes the *causal power* of human action, as people 'make their own history' with the crucial caveat that 'they do not make it as they please; they do not make it under self-selected circumstances, but under circumstances existing already, given and transmitted from the past' (Marx 1852). This 'open Marxism' (Buddharaks 2018) comprises an integrated social science, extending importantly to natural-scientific concerns, as developed for instance by Richard Lewontin and Richard Levins (2007), guiding a vibrant research programme (Burawoy 1990).

Let us, then, glean some insights from that research programme, as they pertain to the phenomenon of Fossil Capitalism.

As Marx (1976a[1867]:125) pointed out in the first sentence of his masterwork, *Capital*, 'the wealth of societies in which the capitalist mode of production prevails appears as an "immense collection of commodities"; the individual commodity appears as its elementary form.' For Marx, analogizing with living systems, the commodity is the cell-form for capitalism as an organic system. Every commodity is composed of two features: exchange-value (what it can fetch when exchanged for other commodities) and use-value (its usefulness to a prospective purchaser). Capitalism, as noted, is 'generalized commodity production', and Marx's analysis of exploitation is pitched at a high level of abstraction, namely capitalism as a whole.

Abstracting from local and temporary conditions (including fluctuations in supply and demand, which influence current prices), Marx asserts that in a market system commodities are exchanged at their values – proportionately to the amount society's total labour-time ('abstract labour') that is needed to produce a given commodity. An automobile is, on this reckoning, more valuable than a loaf of bread because it takes much more of society's total labour-time to produce the former – from the mining of iron (and these days, lithium) to the processing of raw materials, fabrication of parts and final assembly. In generalized commodity production, human agency itself takes on a commodity form, as *labour power* sold by workers to capitalists in exchange for a wage or salary. The value of labour power is the amount of abstract labour needed to produce it, namely, the worker – on a daily and intergenerational basis (the latter supporting the worker's family). This is materialized in the wages that capitalists pay workers. Thus, the exchange of labour-power for wages is a fair one: workers, at this high level of abstraction, receive as wages the full value of their labour power.[2]

Marx's key insight is that labour power, a peculiar commodity, has as its use-value the capacity to create new value that is greater than what the capitalist advances as the wage. The whole point in employing workers is to do this; otherwise the capitalist's other assets – machines, raw material – sit idle. Nothing happens without human agency, and under capitalism, that agency is commodified. Once the worker enters the workplace, his or her labour power belongs to the capitalist who has purchased it and has the right to consume it. This peculiar commodity is consumed 'productively', as concrete labour within various production processes. The labour performed – workers' agency in production – creates not only the value equivalent of the wage, but additional value, surplus value, which forms the basis of profit. The capitalist appropriation of surplus value is at the core of exploitation. As Nancy Fraser (2022:15) summarizes,

> Exploitation transfers value to capital under the guise of a free contractual exchange: in return for the use of their labor power, workers receive wages that (are supposed to) cover their costs of living; while capital appropriates their 'surplus labor time,' it (supposedly) pays at least for their 'necessary labor time.'

As is well known, Marx emphasized the dialectical relation between relations and forces of production as they develop within modern capitalism. These are broad categories that enable us to delineate, on the one hand, the social relations into which people enter as they materially produce and reproduce a specific way of life (i.e., relations of production) and, on the other hand, '*the practices, objects, techniques and knowledges through which we are purposefully linked to and transform the rest of nature*' (i.e., forces of production; Graham 2021:2, emphasis in original). This distinction is important, but it is purely analytical: in lived reality, forces and relations of production interpenetrate and are indispensable to each other. The complex combination of these mutually constitutive relations and forces comprises the capitalist mode of production. Although the capital-labour relation is at its heart, as we have seen with the treadmill of production, capitalists are also interrelated. In competing for shares of total surplus value they are compelled to introduce techniques and technologies that enhance labour productivity. Businesses that improve efficiency in this way gain a temporary advantage by reducing their overall cost of production. They thereby realize higher profits. But competitors will introduce the same innovations as soon as they can, sparking a new round of competitively induced innovation.

In this way, the capitalist drive towards endless growth not only accelerates the treadmill; it tends to develop the forces of production, particularly as science becomes fully integrated into the capital accumulation process. However, and this is key, *it is profit-motivated capitalists who control this process*. Advances in the efficiency of production 'are always contained within capitalism's exploitative relations of production' (Carroll and Sapinski 2018:9), and inter-capitalist competition pushes capitalists to externalize costs by making the withdrawals from and additions to ecosystems that degrade living systems. Productive forces are permeated by productive relations, so that 'progress' is conditioned and framed by the dominant position capital holds in those relations. However, other social interests are also active in developing the productive forces, and we must emphasize that these forces take in not only technology but the skills and knowledge of those who make production happen. Most importantly for our purposes, as Nicolas Graham emphasizes, 'ecological knowledge, including recognition of the need to restore and maintain the indispensable metabolism between humanity and nature, represents an advancement in the productive forces' (2021:3). Examples of these 'green forces of production' include not only ecological knowledge but 'the growth of renewable energy, "green" infrastructures (e.g., low-carbon transportation, energy efficiency measures) and agroecology' (ibid.). Far from being a neutral, benign impulse towards 'progress', there is contention over the form and content of productive forces.

Graham's reconceptualization of productive forces carries an important implication for our understanding of the ecological crisis. In Marx's original formulation, 'at a certain stage of development, the material productive forces of society come into conflict with the existing relations of production.... From forms of development of the productive forces these relations turn into their fetters' (Marx 1976b:21). Graham's thesis is that the emergent forces of production that cluster around ecological knowledge and associated practices such as ecological restoration and agroecology, the green productive forces, 'are fettered by capitalist relations of production' (2021:4). That is, 'while ecological knowledge continues to develop and deepen today (partially as an outgrowth of the ecological contradictions of capitalism), it is *underutilized*, languishing at the margins of an anti-ecological system' (2021:9). It is true that some fractions of the capitalist class and its acolytes now favour 'green growth' or 'clean growth' (to be discussed in Chapter 4), but all schemes that fall under this rubric are designed to provide opportunities for profitable capital accumulation. Moreover, according to Emilano Brancaccio (2023), among a growing fraction of capitalists a quickening current of 'anti-ecological capitalism' holds that 'the green transition is taking place too fast, creating a risk that rising production costs will become unsustainable.' Sustainability, in this context, refers not to the health of ecosystems but to the conditions for profitable (and thus 'sustainable') capital accumulation, which are eroded by the higher production costs necessitated by any serious ecological transition.

We can unravel this strand of analysis further by turning to an influential formulation of ecological Marxism by James O'Connor (1988), namely, the 'second contradiction of capitalism.' For O'Connor, the global ecological crisis 'marks a point in the development of capitalism at which new barriers to capital accumulation and new forms of systemic crisis—in other words a new contradiction—appear' (Spence 2000:85). The contradiction is between the process of capital accumulation (i.e., the continuing development of capitalist relations of production and forces of production) and its own 'conditions of production', which the accumulation process undercuts, creating barriers to further accumulation.

> Examples of capitalist accumulation impairing or destroying capital's own conditions hence threatening its own profits and capacity to produce and accumulate more capital are well-known. The warming of the atmosphere will inevitably destroy people, places, and profits, not to speak of other species life. Acid rain destroys forests and lakes and buildings and profits alike. Salinization of water tables, toxic wastes, soil erosion, etc. impair nature and profitability.
>
> *(O'Connor 1988:22)*

As capital degrades the web of life, the costs and demands of reproducing the conditions of production escalate, undercutting profitable accumulation and often requiring increased state mediation, via direct and indirect subsidies to business.

O'Connor's 'second contradiction' further develops Schnaiberg's insight that capital treats the conditions of production as inexhaustible factors of production and externalizes 'costs' into naturalized realms (Rudy 2019). In Schnaiberg's analysis, as production becomes more capital-intensive the treadmill must accelerate to avoid endemic unemployment (and the economic stagnation it begets), yet that acceleration further degrades ecosystems. In O'Connor's analysis, such degradation over time undermines profitability and therefore accumulation itself, threatening economic stagnation. To mitigate this crisis tendency and to manage the damage inflicted upon ecosystems capitalist elites have turned to 'sustainable degradation' (framed for public consumption as 'sustainable development'), which in Timothy Luke's view 'is a proactive, profitable and powerful policy that maintains some environmental viability by creating zones and spheres of control where degradation is lessened, but never stopped' (2006:101).

Of course, the conditions of production are not an inexhaustible collection of 'natural resources'; they are elements of living systems, including human labour itself. Jason Moore (2016) has argued that the result of capitalism's epochal global expansion is a sharp contraction of potential sources of the 'cheap nature' that has been integral to capitalist profits. For instance,

> Since the 1970s, low-cost frontiers of oil extraction – in Alaska, the Gulf of Mexico, West Africa and the North Sea – have been superseded by high-cost frontiers in Northern Canada. There, plentiful tar sand deposits are carbon intensive and expensive to refine. Meanwhile, there are no new land frontiers on which to grow cheap food, just as global warming undermines the existing capacities of livestock farming and crop growth.
>
> *(Hope 2018:570)*

Later in this chapter, I will outline how capitalism's reliance on cheap natures has been at the centre of practices of colonialism and imperialism, to source cheap labour and raw materials, wreaking ecological and social havoc upon communities of the global South. At this point, I turn to one further formulation that builds directly on Marx in illuminating another basic facet of capitalism's inbuilt tendency towards ecocide – the theory of metabolic rift, as elaborated by John Bellamy Foster and his colleagues (Clark and York 2005; Foster 2022).

Unlike most social scientists, who embraced a dualism between humanity and nature, Marx and Engels were keen students of natural science, and they integrated natural-science insights into their analysis of human society. 'Marxian materialist dialectics was grounded in human corporeal existence within the physical world, in a context of emergence, or integrated levels' (Foster 2022). Drawing on the breakthroughs in soil science achieved by Justus von Liebig, Marx introduced the concept of metabolic rift in his critical analysis of the ecological problems that stemmed from industrialization, urbanization and the reorganization of agriculture within 19th century capitalism.

In the 1850s, Liebig had shown how agriculture in England and the United States had become a 'spoliation system' that, in removing nutrients from the soil as produce was shipped to urban centres undermined the 'conditions of reproduction' of the soil (Liebig, quoted in Foster 1999:378). In *Capital*, Marx wrote:

> Capitalist production collects the population together in great centres, and causes the urban population to achieve an ever-growing preponderance. This has two results. On the one hand it concentrates the historical motive force of society; on the other hand, it disturbs the metabolic interaction between man and the earth, i.e. it prevents the return to the soil of its constituent elements consumed by man in the form of food and clothing; hence it hinders the operation of the eternal natural condition for the lasting fertility of the soil. … All progress in capitalist agriculture is a progress

> in the art, not only of robbing the worker, but of robbing the soil; all progress in increasing the fertility of the soil for a given time is a progress toward ruining the more long-lasting sources of that fertility. ... Capitalist production, therefore, only develops the techniques and the degree of combination of the social process of production by simultaneously undermining the original sources of all wealth—the soil and the worker.
>
> *(Marx 1976a [1867]:637–38)*

In this passage we can discern a recognition that 'progress' in the forces of production as they develop within the logic of capital (as in the introduction of fertilizer to compensate for 'robbing the soil') can have pernicious implications. Liebig himself had argued a few years earlier that "rational agriculture, in contrast to the spoliation system of farming, is based on the principle of restitution; by giving back to the fields the conditions of their fertility, the farmer insures the permanence of the latter" (1859:183, quoted in Foster 1999:378). This position, which Marx accepted, anticipates contemporary principles of agroecology, representing green forces of production. The passage also illustrates O'Connor's second contradiction, as the conditions of production – both nonhuman (soil) and human (the worker) – are degraded by the accumulation process.

Foster's important contribution was to extend Marx's germinal analysis beyond the agricultural sector. In a recent interview, he explained,

> The basic idea of the metabolic rift is not very difficult. The human relation to nature, like that of all life, is a metabolic one, that is we appropriate energy and material resources from the environment as a basis for life, metabolize this in our bodies, and return the waste to the earth. In the case of human beings, as the self-mediating beings of nature, our relation to nature takes the form of a social metabolism exercised primarily through the labor and production process. However, with the development of capitalism, this social metabolism was alienated: humanity became more estranged from the earth, as evident in what Marx called the 'original expropriation,' or the removal of populations from the land in the fifteenth through nineteenth centuries, and the expropriation of land, resources, and human bodies throughout the globe, forming the basis of industrial capitalism. Nature in this system is no longer seen as a relation to which we belong, but something to be conquered and treated as a 'free gift' to capital.
>
> *(Foster and Sarican 2023)*

Brett Clark and Richard York specifically applied the concept of metabolic rift, understood as 'the rupture or interruption of a natural system' (2005:400), to the issue of climate change, to which I now turn.

Clark and York begin their analysis at the very highest level, that of the earth system and the emergence of life within it. It is worthwhile, I think, to walk us through the steps in their argument. 'Life – in interaction with the existing environment – created the atmosphere, as we know it. Life exists only in the lower regions of the sky and upper regions of the soil and ocean. An interrelationship between living and nonliving materials within the biosphere produces a cycling of chemical elements' (2005:400) – including the nutrient cycle that is disrupted by capitalist agriculture's 'spoilation system' and the carbon cycle, whose disruption is the main driver of climate change. Attending to the latter cycle, Clark and York continue:

> Life on earth depends upon energy from the sun for its existence. The sun's energy is captured by plants, which store and convert it into chemical energy for its own growth. At the same time, animals eat plants to derive the necessary energy for their lives. Through plants and animals, energy is captured, stored, converted, and deposited throughout the environment, maintaining a viable world for life and its evolutionary processes. Fossil fuels hidden deep within the earth are the remains of past life, especially the first wave of gigantic ferns and giant trees.
>
> *(401)*

The carbon cycle itself takes in the entire biosphere, 'as carbon moves through the air, rocks, soil, water, and all living things in a cyclical process,' upon which all life depends. Some carbon is incorporated into nonliving entities – oceans, glaciers and the like – which function as 'sinks', limiting the concentration of CO_2 in the atmosphere. Atmospheric carbon also becomes fixed within plants as they synthesize it with solar energy, water and chlorophyll to produce carbohydrates and oxygen. In turn, 'much of that carbon is passed on to other species, and onward through the food chain, where carbon enters the soil and water as waste, as dead matter, or as CO_2 through the respiration of animals.' This cycle, in which 'CO_2 is released into the atmosphere only to be recirculated to the earth through a variety of pathways in natural processes' is integral to all life (402).

Capitalism emerged between the fifteenth and nineteenth centuries, in a process of 'primitive accumulation' that established its historical premises: wage labour, capital and conversion of land (in effect, nonhuman nature) into commodity form. In this epochal process, one in which, as Marx (1976a[1867]:926) famously described, 'capital comes dripping from head to toe, from every pore, with blood and dirt,' Europe's peasants were dispossessed of the commons, pushed from the countryside to emerging urban centres and pressed into service as proletarians in capitalist workshops. As the other side of the same process, capitalists, previously on the margins of feudal society as merchants and money-lenders, began to gain control of the productive forces, in the form of small-scale firms and commercial farms employing a few hands. These developments primarily in Europe were matched by a dramatic expansion of European colonization. The major European powers – Spain, Portugal, the Netherlands, England and France – seized control of what we now call the global South. Under colonial control, people were dispossessed of their land and forced to work (often literally as slaves) on plantations and in mines for the colonizers, who also stole precious artefacts, gold and silver, which provided an expanding monetary basis for capital accumulation.

At this point, as Clark and York continue, capitalism's development of the forces of production is important. At first, production relied on energy from human beings as they used tools in the production of commodities. Early capitalism, as J.R. McNeill (2000) has observed, was a 'somatic' regime, mobilizing energy from food sources through human (and animal) bodies, to muscle power. As the scale of production grew, the limitations of human- (and animal-) powered manufacturing began to pose a barrier to accumulation, impelling capitalists to harness other forms of energy for commodity production. Thus, 'the movement from human motive power to water and wind to coal-driven steam-engines transformed capitalist production, increasing the scale of production by pushing up labor productivity to historically unprecedented levels, and by deepening the exploitation of nature and labor' (Clark and York 2005:405). Fortuitously, James Watts's refinement of the coal-powered steam engine, patented in 1769, gave capitalists the means to establish 'the world's first "exosomatic" regime: the first to take carbonized solar energy from beneath the crust of the earth and convert it to mechanical energy *outside of living bodies*' (Fraser 2022:95; cf. Pineault 2023:73–89). The age of 'Fossil Capitalism' had dawned.

Fossil Capitalism and Fossil Capital

According to Smil's (2010:155) calculations, in 1800 coal accounted for only 1.7 percent of global energy consumption (most of which involved the burning of wood and peat). By 1867, when Marx published the first edition of *Capital*, coal's share had risen to nearly 20 percent. Coal was the leading form of global energy consumption by 1910, claiming 55.4 percent, but of course subsequently petroleum overtook coal, so that by 2008 crude oil's share was 30.9 percent, coal's share was 28.9 percent, with natural gas claiming a further 22.8 percent. By these calculations (which are based on exajoules as a measure of energy), fossil fuels comprised 82.6 percent of global energy consumption in 2008. In 2022, according to the *Statistical Review of World Energy* (Energy Institute 2023:3), 'fossil fuel consumption as a percentage of primary energy remained steady at 82 percent' – compared to a 7.5 percent share for all renewables, including hydroelectricity.

Across two-and-a-half centuries, the development of industrial capitalism has closely tracked the growth of fossil fuel usage, and global capitalism continues to be powered by fossil fuels. As Elmer Altvater stated, in the essay that first introduced the term 'fossil capitalism', 'at the centre of the analysis of capitalism's relation to nature is its inherent and unavoidable dependence on fossil fuels' (2006:39). This is because of the very high 'Energy Return on Energy Input (EROEI), particularly in the case of oil: 'only a small amount of energy needs to be invested in order to harvest much greater amounts of energy, because the entropy of petroleum is very low and its energy concentration is very high, yielding a high energy surplus' (2006:39). However, EROEI falls as the most readily available sources of oil are exhausted, requiring an increasing amount of energy to be invested in extraction. Yet fossil energy remains entrenched at the core of capitalism's system of production. For Altvater, 'One of the main advantages of fossil energy for capitalist accumulation is the *congruence* of its physical properties with the socioeconomic and political logics of capitalist development' (2006:41). The transition to capitalism itself

> was a revolutionary break in the history of the societal relation of human beings to nature because it was no longer the *flow* of solar radiation which served as the main energy supply for the system of production and the satisfaction of human needs, but the use of the mineralised *stocks* of energy contained in the crust of the earth.
>
> *(Altvater 2006:40)*

Altvater outlined several reasons for the close fit between capital accumulation and fossil energy. The latter enabled the transformation from localized production and consumption to the capitalist pattern in which energy resources are transported to industrial sites anywhere in the world. It detached the energy system from natural rhythms (day/night, seasons), since 'fossil energies can be stored and then consumed without reference to natural time patterns, in accordance only with the time regime of modernity and a timetable that optimizes profits' (2006:41). Moreover, fossil energy can be incorporated into production, transportation and consumption very flexibly, empowering management to follow the logic of profitability without having to take energy restrictions into account.[3] In the long term this logic has rendered capital accumulation 'increasingly independent of natural conditions and their limitations' (2006:41–2).

If Altvater's pathbreaking intervention showed the perfect congruence of 'capitalism, fossil energy, rationalism and industrialism' (2006:42), a decade later Andreas Malm's *Fossil Capital* provided an historical analysis and theorization of the emergence and development of Fossil Capitalism. Malm's 488-page, highly readable tome offers two key insights that build on Altvater.

First, given fossil capital's primacy in production, capital's drive to endless growth, means increasing quantities of fossil fuels burned, resulting in increasing carbon pollution. Malm argues that

> at a certain stage in the historical development of capital, fossil fuels become a necessary material substratum for the production of surplus-value. ... Other sources of mechanical energy are pushed to the fringes, while capital expands in leaps and bounds, energised by fossil fuels. These have now become the general lever for surplus-value production.
>
> *(2016:288)*

As the form of capital that energizes capitalism, fossil capital is 'self-expanding value passing through the metamorphosis of fossil fuels into CO_2' (2016:290).

Second, Malm places capitalism's relations of production at the centre of this history. In contrast to accounts that depict the shift from water power to steam power in the first few decades of 19th century Britain as resulting from purely technocratic calculations, Malm notes that the water-wheels of early capitalism continued to produce energy at lower cost than coal. The shift was driven by strategic calculations by capitalists engaged in competition with each other yet united in class struggle against an incipient labour movement.

On the one hand, as competitors, capitalists 'have no reason to share plans for production ex ante; competition throws a spanner in the works of mutual adjustment, blocks the sharing of information and upsets collective plans' (Malm 2016:296). Although run-of-the-river water power was cheaper than coal-fuelled steam, the former was a common-pool resource, requiring cooperation, planning and trust among participants – as is the case with any element of the commons. The unwillingness, or inability, of mill-owners 'to submit to the planning, coordination and collective funding required for expansion of waterpower capacity' (2016:118) relegated water power to the margins, as the scale of the economy expanded. As buried sunshine, a relic concentrating past solar energy into a mobile substance that could be fully commodified (thereby skirting the need for planning and cooperation) coal fit exactly with capital's atomizing profile. Above all, 'it conferred on the capitalist the freedom to store energy and mobilize it at the desired moment and in the degree needed' (Fressoz-Bonneuil 2017:55).

On the other hand, and concomitantly, class struggle provided a strong impetus for the adoption of coal as industrial capitalism's primary energy source. Within production, capital's power over labour is conditional upon control over energy as the prime mover that 'makes everything work' (Malm 2016:314). Unlike the flow of water, as a stock of commodified energy, coal could be fully commodified and brought within the circuitry of accumulation: purchased and transported to capital's increasingly centralized dark satanic mills. This not only provided a reliable energy source, it weakened traditional, community-based resistance, as atomized proletarians uprooted from rural communities faced off against the employers on whom their subsistence depended.

> With the stock as a power at their command, capitalists inflated their power vis-à-vis workers.... By dint of their exceptional purchasing power, capitalists could buy steam engines and coal alongside slices of human lives over which they could then exercise reinforced power on the shop floor...
>
> *(Malm 2016:315)*

Citing documents from the Chartist movement, Malm (2016:223–48) traces the working-class resistance to the new forces of production, which climaxed in the general strike of 1842. But as capital inexorably penetrated (and created) more lines of production, and particularly with the rise of coal-fired 'general manufacturing' around mid-century, Britain consummated the world's first fossil economy (2016:250), which became the model for industrialization everywhere.

Jump-cutting to our current setting, Nicolas Graham and I have summarized the upshot, and one fateful ecological implication:

> In little more than two centuries fossil capital has become embedded in the entire economic structure, from industrial production to plastics and petrochemicals, to the agro-industry, to transportation, and all else. New oil and gas projects and pipelines add to this mass of infrastructure, 'locking-in' carbon in an era of deepening climate crisis.
>
> *(Graham and Carroll 2022:27)*

Imperialism and Fossil Capitalism

So far, our discussion has offered an ecological take on Fossil Capitalism as an industrial economy of generalized commodity production, powered by fossil fuels and centred upon the relationship between capital and wage labour. But as Nancy Fraser (2022), among others, has pointed out, capitalism is more than an economy; it is a way of life. James O'Connor's (1988) emphasis on the 'conditions of production' gives us some purchase on how capitalism is reproduced as a form of human society. Although O'Connor's 1988 article was taken up primarily for its insights on ecological Marxism, the scope of O'Connor's intervention was broader (Spence 2000). O'Connor based his analysis on Marx's delineation of three conditions of production for capital: (1) the 'external physical conditions' (the natural elements entering into production), (2) the 'personal conditions of production' (which, detached from the workplace, reproduce human beings as wage-workers with appropriate capacities and motivations) and (3) the 'communal, general conditions of social production' (as in infrastructure for production and 'means of communication') (O'Connor 1988:16).

In her discussion of 'cannibal capitalism', Fraser has extended the concept of conditions of production to identify the 'vital underpinnings of accumulation' that 'are constitutive components of the capitalist order' (2022:xiv). Although these supports are indeed essential, capital *expropriates* the wealth they represent without (full) compensation. Cannibalistic practices, which amount to theft of labour and nonhuman nature, enable and enlarge profits, but as O'Connor observed, as capital free-rides on its own conditions of production, it degrades them (thus Fraser's invocation of cannibalism). In the making of Fossil Capitalism, a particularly significant extra-economic support has been 'the forcible seizure, on a continuing basis, of the wealth of subjugated and minoritized peoples' (Fraser 2022:14) – a process of colonialism and, later, imperialism, in which capital, radiating from what we now call the global North, in collaboration with strong states, has dominated the so-called global South.

In contrast to the economy of 'free wage labour' – wherein workers, although exploited, are accorded basic rights as owners of property (namely, their labour power and personal effects), in expropriation

> capitalists dispense with all such niceties in favor of brute confiscation of others' assets, for which they pay little or nothing; by funneling commandeered labor, land, minerals, and/or energy into their firms' operations, they lower their production costs and raise their profits. Thus, far from excluding one another, expropriation and exploitation work hand in hand. Doubly free wage laborers transform looted 'raw materials' on machines powered by confiscated sources of energy. Their wages are kept low by the availability of food grown on stolen lands by indebted peons and of consumer goods produced in sweatshops by unfree or dependent 'others,' whose own reproduction costs are not fully remunerated.
>
> *(Fraser 2022:15)*

Within historical materialism, imperialism has been a particularly important concept in understanding how, in the era of monopoly capital, businesses and states have been entangled in geopolitical relations of expropriation and domination (Brewer 1980; Hobsbawm 1989; Van der Pijl 1998). In this section I unfold the concept of imperialism and consider its entanglement with fossil capital.

'Imperialism' carries multiple meanings; often it is employed descriptively, a synonym for empire. Within historical materialism, imperialism highlights the *uneven development of Fossil Capitalism on a global scale* from the late nineteenth century forward and the *role of advanced capitalist states in shaping global capitalism to suit the interests of big, internationalizing capital.* Broadly defined, imperialism amounts to 'the struggle of large, monopolistic capital over economic territory, actively aided and assisted by states' (Ghosh et al. 2022:70).

I begin from John Weeks's concise formulation, which presents a framework for comprehending imperialism as monopoly capitalism operating in a global field. Weeks writes,

> We live in a world dominated by capitalism. As a consequence of this fact (over which there is presumably no controversy), the theory of the operation of the world economy must have a theory of capitalist reproduction as its basis. By 'capitalist reproduction' is meant the process by which a specifically capitalist society evolves and reproduces its social relations on an expanding scale. Second, a theory of the operation of the world economy must locate itself internationally. By this we mean that one needs to derive within the theory a distinction between the 'domestic' reproduction of capitalist society and its 'international' reproduction.
>
> *(1981:118)*

On the first of these, Weeks recapitulates the classic Marxist analysis, showing how the accumulation of capital reproduces its premises (wage labour and capital) on an expanding scale, developing productive forces under capitalist control. On the second issue,

> Materialist theory converts its theory of accumulation into a theory of the world economy by locating it explicitly in the context of countries. What makes a political territory a 'country' is that the territory is controlled by a distinct ruling class, the vehicle for such rule being the state.
>
> *(Weeks 1981:121)*

In this elucidation imperialism amounts to the accumulation of capital in the international context of state-mediated class struggles and uneven development. The relations comprising imperialism are of three kinds. There are conflicts (and cooperation) among the ruling classes of advanced capitalist countries. There are conflicts (and cooperation) between advanced capitalist ruling classes and the ruling classes of peripheral countries (the 'articulation' of modes of production). And, there are conflicts between ruling classes and 'oppressed peoples' (ibid.) – those 'dependent or unfree subjects, enslaved or colonized, unable to call on state protection and stripped of every means of self-defense' (Fraser 2022:89).

'Imperialism' gives us purchase on the geopolitical economy of capital. Recent contributions have developed the ecological side of this phenomenon. As with the colonialism that was essential to capitalism's rise in the

15th through 19th centuries, imperialism has, particularly in its North-South relations of expropriation, relied on what David Harvey terms *accumulation by dispossession* (Harvey 2003). Put simply (and for the moment bracketing off the issue of inter-imperial relations), 'imperialism is the system by which a dominant power is able to control the trade, investment, labor, and natural resources of other peoples' (Frame 2022:508). The global expansion of capitalism from its western-European epicentre, despite selective industrialization of the global South beginning in the 20th century's latter decades, has reproduced a North-South structure of domination in which

> core countries remain characterized by high development of forces of production, including high-paid labor, capital-intensive production, and technological advancement. In contrast, peripheral countries are characterized by low-level forces of production, including low-paid labor and labor-intensive production, such as the export of raw materials.
>
> *(Frame 2022:511)*

For Frame, ecological imperialism results in '*negative socio-ecological impacts or ecological debt or ecologically unequal exchange for peripheral countries, which cannot be truly compensated in a monetary sense*' (2022:509). In ecological unequal exchange, transactions between the advanced capitalist core economies and the extractive periphery eventuate in a deterioration in the ecological situation on capitalism's periphery as expropriation drives resource consumption, accumulation and technological advances in the advanced capitalist North (Frame 2022:507; cf. Buller 2022:221). Frame concludes that 'ecological imperialism has necessitated a wholesale reconfiguring of socio-ecological relations between humans and nature and the subjugation of the political, economic, and institutional structures in the Global South' (2022:525).

Extractivism – conventionally defined as extraction of enormous volumes of natural resources, at most semi-processed and mainly for export, according to the demand of the imperialist countries of global capitalism's core (Gago and Mezzadra 2017:576) – has been central to ecological imperialism, as a growing literature attests (Bowles and Veltmeyer 2017; Arsel and Pellegrini 2022; Ghosh et al. 2022; Dunlap n.d.). But as a basic aspect of ecological imperialism extractivism 'also targets the labor and life' of populations on capitalism's periphery, extracting value from them through expropriation. Moreover, in funding extractive operations, Northern-based finance exercises 'a command over the future', integrating the field of extractive production through the logic of financial capital (Gago and Mezzadra 2017:579–83), and locking many counties on capitalism's periphery into extractive capitalism, with all its economic and social maladies.

Given that modern capitalism has been powered by buried sunshine, it is not surprising that, across the epoch of imperialism, extracting carbon, and particularly oil, has been central in the global accumulation process. As I have argued elsewhere,

> Eras of imperialism have been interdependent with epochal shifts in fossil capitalism.... Broadly, we can distinguish an emergent phase, during Pax Britannia (covering the latter half of the nineteenth century), a classic phase from the early twentieth century to 1945, a post-war phase termed Pax Americana (Cox 1987), and a post-1970s phase as American hegemony gave way to 'global governance', partially integrating high-growth Southern countries (notably, the BRICS). During each phase, fossil capital has played a pivotal role within a changing global political economy and ecology.
>
> *(Carroll 2020:32) (see Table 1.1)*

In the first era, which established modern imperialism's preconditions, coal was 'king' in powering the transition from competitive capitalism organized via national markets to monopoly capitalism, which came to outgrow those

TABLE 1.1 Historical eras of imperialism and fossil capitalism

Era	*Geopolitical economy*	*Political ecology of fossil capital*
Pax Britannia Mid- to late 19th century	Free trade, decline of competitive capitalism, (continuing) colonization of the South	King Coal; steam-powered industrialization and transport
Classic imperialism 20th century to 1945	Great-power rivalry for control of colonies, resources and markets, development of monopoly capitalism	Rise of Big Oil; Seven Sisters develop oilfields in the Middle East, Sumatra, Mexico etc.; emergence of automobility
Pax Americana Post-World War 2 to 1970s	US Open Door policy within Cold War, transnational corporations, Fordism and imperial mode of living, (IML) neocolonialism and national liberation	Carbon democracy; full-throttle automobility; oil becomes lifeblood of consumer capitalism; burgeoning CO_2 emissions
Neoliberal global governance Late 1970s – today	Decline of US hegemony, neoliberal globalization via 'global governance' and spread of IML, rise of BRICS and global South, current realignments	Gathering climate crisis; increasing resort to extreme oil; emergence of 'Climate Capitalism'

Adapted from Carroll (2020:33).

markets. Under British leadership, this period saw northern-based capital establish coalmines from Tasmania to Trinidad and elsewhere, as a world market was consolidated alongside the deepening of colonization.

In the era of classic imperialism the petroleum industry was consolidated in a few giant firms (the so-called 'seven sisters'); indeed, across various industries big corporations and financial institutions came to dominate the economies of the North, controlling enormous economic surpluses that could not be absorbed by domestic markets. As monopoly capital internationalized, the world became divided among the great powers, 'engendering pitched rivalries over control of colonies (including the oil-rich Middle East) as zones of cheap labour, resource extraction and capital export, and leading to two world wars' (Carroll 2020:33). In this era, 'world development was characterized by the confrontation between imperialistic blocs and national capitals that they represented' (Bonanno 2000:319).

The post-World War 2 era of Pax Americana is known to climate scientists as the era of the 'Great Acceleration' in atmospheric carbon emissions (McNeill and Engelke 2014); indeed, most of the greenhouse gases causing the planet to warm have been sent into the atmosphere since the late 1940s. This was a time of unprecedented economic expansion, powered increasingly by oil and stoked initially by the need to rebuild from massive wartime destruction. The immediate post-war years presented a boon particularly to US-based capitalists (who had of course profited handsomely from the war). As colonial empires were superseded by the US Open Door policy and emergent international organizations (United Nations, the General Agreement on Tariffs and Trade [forerunner to the World Trade Organization]), the United States replaced Britain as the hegemonic leader of a now-unified imperialist bloc.

> If the first part of the twentieth century seemed destined for endless inter-imperial conflict over the spoils of colonialism, by the 1950s a new, collective imperialism was in formation. Increasingly, global trade agreements and lending infrastructures engineered by the old colonial powers would see the spoils of imperial extraction shared among them.
>
> *(Wargan 2023:21)*

Rivalries ebbed and collective imperialism, through the North Atlantic Treaty Organization (NATO) and other platforms, became the order of the day. The chapter following this one focusses on this fateful era, which also saw the flourishing of consumer capitalism and the politics of class compromise in the North, as Fossil Capitalism penetrated 'every aspect of life and every part of the globe' (Angus 2016:152), spurring a sharp increase in carbon emissions.

In imperialism's fourth era, the climate crisis really began to bite. But as Charles Derber (2010) has emphasized, the process was incremental and barely noticeable, akin to a frog being gradually cooked in a pot. Although James Hanson sounded the alarm about the causal link between burning carbon and heating the planet in a presentation before the US Congress in 1988, the issue received no traction until the Kyoto accord in 1997 (which the United States refused to ratify), and to this day the policy response (to be reviewed in Chapter 4) has been insipid.

The explosive growth of fossil capital and of carbon emissions during the Great Acceleration only continued in imperialism's fourth era, which I take up in Chapter 3. For now, we can discern several shifts in this period within the imperialist order, which take us to our current predicament.

In the 1970s, the bloom came off the post-war boom. The long wave of high profits and rapid accumulation ended in stagflation and rising state deficits, and the United States' paramount position as hegemon of the imperialist bloc began to erode. The response in the 1980s, led by the Thatcher regime in London and the Reagan regime in Washington, was neoliberalism, a policy framework that deregulated markets, privatized public assets and attacked organized labour – all with the objective of boosting profits for capital and tearing down remaining barriers to a 'borderless world' for capitalist accumulation. The collapse of Soviet state socialism in 1991 was a great victory for imperialism, and it ushered in a 'uni-polar world' in which American power appeared unchallengeable. In exalting markets neoliberal policy innervated financialization. Increasingly mobile capital flooded into the 'emerging markets' of the global South, abandoning industrial heartlands of the North – particularly in the United States.

Like the post-war 'golden years' of rising prosperity in the North, the unipolar world was a temporary arrangement through which the capitalist juggernaut could move forward. By the second decade of the 21st century, with the rise of the BRICS, one could speak not only of the decline of US hegemony, but the decline of the 'West.' A measure of this was the rapid growth of 'resource nationalism via national oil companies (NOCs), most of them based in the global South. These state-controlled enterprises came to control 77 percent of oil and gas reserves by 2005 (in 1949 the Seven Sisters had control of 88 percent of the global oil trade) (World Energy Council 2016:19). Tectonic shifts were underway in Fossil Capitalism's geopolitical economy and political ecology, but carbon emissions continued to rise, even as the symptoms of climate crisis proliferated.

My earlier discussion of John Weeks's formulation of imperialism identified its three constitutive relations. Scholars of ecological imperialism focus on the compliant relation between the advanced capitalist powers and the ruling elites of global South, which also entails the imperialist relation between ruling classes (both North and South) and the oppression of most of humanity; namely the masses of the global South. But much has changed since the early 20th century. Most tellingly, perhaps, 'for the first time in capitalism's long history, the global economic center of gravity is shifting decisively eastward' (Wargan 2023:16). Indeed, in the new international division of labour that was consummated through neoliberal globalization the centre of gravity for capitalism's working class has also been shifting away from the imperialist core. In the circumstances, Pawel Wargan addresses the question, where do we find imperialism today? His answer:

> We find it among the two billion people who struggle to eat. We find it in the fragility, conflict, or violence that two-thirds of humanity will face in the coming decade. We find it in the many livelihoods that are regularly swept away by rising tides, drought-stricken fields, and creeping desert sands, and among the billion people who do not own a single pair of shoes. ... We find it in the gold and cobalt, diamonds and tin, phosphates and oil, zinc and manganese, uranium and land whose expropriation sees the headquarters of Western corporations and financial institutions grow to increasingly dazzling proportions.
>
> *(2023:21)*

Yet the story of imperialism is also one of resistance by those on its receiving end. In responding to such resistance, imperialism removes its velvet glove. Already in 1917, the success of the Russian Revolution brought forward another facet of imperialism: 'the containment of socialism worldwide' (Carroll 2020:33). In the very first 'coalition of the willing', Britain, France, the United States, Italy and Japan used military force in a vain attempt to defeat the Bolshevik regime. Clearly, the militarism baked into imperialism has had two faces – that of *inter-imperial rivalry* (both world wars of the 20th century) and that of *repressing movements of national liberation* – whether in India, Indonesia and China, in Korea, Guatemala, Iran, Cuba, the Congo, Algeria and Viet Nam, in Chile, Nicaragua and Grenada, or in Afghanistan, Iraq, Libya, Syria and Palestine. Sadly, the list goes on, and the imperial interest in maintaining control of oil and other resources can be inferred rather directly from the countries just listed. Sometimes, repression of anti-imperialist forces has been undertaken by authoritarian regimes, backed by the United States and its allies; at other times, less obviously violent 'regime change' has been engineered through the clandestine operations of the CIA.[4] The imperialist powers have typically disciplined successful national liberation movements to renounce anti-capitalist visions, as in South Africa after the overthrow of apartheid. The few regimes that hold fast to anti-capitalism, such as Cuba, are punished with crippling sanctions, making it clear to the world that attempts to break from the imperial order carry a very heavy price.

Finally, there is the first relation of imperialism Weeks (1981) identified – that of rivalry, *but also cooperation*, among the imperialist powers. If under post-World War 2 American hegemony, the inter-imperialist rivalries that had provoked world wars went into abeyance, they never entirely disappeared. As the American empire has declined, these rivalries have sharpened. Already in the 1980s, as Japan-based capital claimed an increasing share of the American market, the United States required Japan to revalue the Yen, destroying Japan's competitive advantage (and leading to years of economic stagnation in Japan (Beeson and Broome 2010)). In the 2000s, consolidation of the European Union sharpened trans-Atlantic rivalries (Tabak 2015). The 2022 destruction of Europe's Nord Stream Pipelines which has seriously weakened Germany's industrial base and was likely executed by clandestine US forces (Hersh 2023), fits this pattern. Since the close of World War 2, however, the collective imperialist interest, shared by the advanced capitalist states, in maintaining global dominance has far outweighed the tendency towards rivalry.

Yet there is a growing rivalry between the American-led 'collective west' and states outside that imperialist bloc. Indeed, the recent rise of the BRICS (and the enlargement in 2024 to BRICS+), in the context of US-supported wars that appear to be never-ending (Afghanistan, Iraq, Libya, Syria, Ukraine, Palestine, Yemen...), may portend a shift in the fortunes of anti-imperialist movements. BRICS+ comprises a very diverse coalition, and prognoses on its trajectory vis-à-vis imperialism vary, with some Marxists hewing to an interpretation that they are simply 'sub-imperialist' to the US-led bloc – in the same category, say, as Israel (sometimes described as America's 'unsinkable aircraft carrier') (e.g., Robinson 2015; Bond and Fuentes 2024). Be that as it may, it is clear that the impunity with which US-led imperialism has dominated since the 1990s has reached its limits, that the US-led order is in crisis and that the BRICS+ countries are, among states, sites of resistance to that order, opening space for non-state movements. The implications of that for the climate crisis are, to say the least, complex.

Hegemony: Organizing Consent to Fossil Capitalism

Complementing the concepts of capital accumulation and imperialism, the third core concept in this book's theoretical and historical analysis is hegemony. Simply put, hegemony is rule with consent of the ruled, a form of *leadership* in which persuasion predominates over coercion, but is backed up by the threat of coercion. Combined with accumulation and imperialism, the struggle for hegemony has been a fundamental process in the development of Fossil Capitalism. Our discussion of the leading role Britain and the United States played in subtending the capitalist world order from the mid-19th century through the 20th century has called attention to the role of hegemony in reproducing an expanding capitalism within a global field. Yet the struggle for hegemony is multi-scalar, involving international, national and subnational organizations and practices right down to local everyday life, and extending well beyond the state as conventionally understood.

Like imperialism, the concept of hegemony points in the direction of an extra-economic support that is integral to Fossil Capitalism. But if imperialism is about subjugating (neo)colonies and extracting wealth through dispossession in a geopolitics of uneven development, hegemony is about leadership and the persuasive power of ideology. The concept was developed by socialist activist Antonio Gramsci (1891–1937), whose arrest and incarceration by the fascist Mussolini regime in 1926 barred him from functioning as General Secretary of the Communist Party he co-founded, restricting his activities to reading and writing. By 1935, when deteriorating health required his removal to a clinic under police guard, Gramsci had composed, in the *Prison Notebooks*, a brilliant elaboration of historical materialism filling 3,369 pages in the critical edition of 1975.[5] The theoretical focus of Gramsci's interest was the state. Gramsci viewed the state as a site not only of sovereign, coercive power (the legislature and executives that create law and the courts, police, and military that enforce it, protecting private property and the broader capitalist order) but of legitimation, enabling the reproduction of capitalism with the consent of the subaltern populations whose agency and wealth it exploits and expropriates.

Indeed, the question of the state reopens the issue of reproduction. Unlike previous ways of life in which economic and political power were fused (e.g., European feudalism), within capitalism these powers are split apart (Wood 1981:81). The power to exploit and expropriate resides in a separate realm, the market economy. Political power is lodged in another realm: the state – with the paradoxical implication that 'the ruling class does not rule' (Block 1977), at least not directly (Carroll and Sapinski 2018:17). Instead of capital's direct rule, the structural dependence of the state on capital as the ultimate source of its own revenue and the many tentacles of influence that extend from capital's private power to the state's public power severely limit the scope of political discussion and policy. Charged with the task of managing public order to maintain capitalism's premises – private property, a compliant workforce available for hire, and conditions for profitable accumulation – the capitalist state's democratic capacity is highly constrained. Capital's sovereign control of economic life

> deprives us of the ability to decide collectively what and how much we want to produce, on what energic basis and through what kinds of social relations. It deprives us, too, of the capacity to determine how we want to use the social surplus we collectively produce; how we want to relate to nature and to future generations; how we want to organize the work of social reproduction and its relation to that of production. By virtue of its inherent structure, then, capitalism is fundamentally anti-democratic.
> *(Fraser 2022:122)*

Within capitalist democracies those managing state power must mediate the contradictory relation between an anti-democratic economy and a state apparatus whose top tier is elected by the citizenry. The monopoly form of late capitalism concentrates both economic and state power within a 'power bloc' as economic and political leaders encompass a '*partnership between two different, separate forces*, linked to each other by many threads, yet each with their own sphere of concerns' (Miliband 1983:65).

Such mediation, within a partnership, requires the efforts of legions of 'organic intellectuals', 'the thinking and organising element' that promulgates and directs 'the ideas and aspirations of the class to which they organically belong' (Hoare and Smith 1971:3). Such people (including politically active capitalists) are the 'deputies (specialised employees)' entrusted with 'organising the general system of relationships' external to capitalist enterprises themselves (Gramsci 1971:6). Politically active capitalists serve this function as 'business leaders', but most organic intellectuals are 'specialised employees', professionals socialized into world views that identify markets and private property with equality and freedom. Their ranks include politicians and senior civil servants, lobbyists, industrial technicians, liberal economists, accountants, lawyers, consultants, mainstream journalists, 'the managers and minions of the culture industries' and the like (Carroll 2024:10). Through their varied initiatives in the political and cultural fields, they create the policies and cultural texts that shape our world, recruiting loyalty to that order. Hence,

> the production of consent relies on 'organic intellectuals', who function as persuaders by propagating particular ideas. They provide legitimation for the dominant group's economic, political, cultural, moral and intellectual leadership...
>
> *(Muhr 2024:408)*

Thomas Muhr goes on to note, however, that 'hegemonic struggle involves each social group having or forming their own organic intellectuals' (ibid.). Ruling-class hegemony is contested, and in that struggle activists with the knowledge and skills to seek an alternative – a liveable world for all – are formed. Contemporary initiatives to reform or transcend Fossil Capitalism, which I take up in part 2 of this book, are testimony to that.

Successful hegemony is evident in 'the ability of the ruling classes to enforce their interests in such a way that they are seen as the general interest by the subaltern classes' (Brand and Wissen 2024:279). Defining, and enforcing, the 'general interest', at the level of national states, or the international interest at a global level, is a crucial element in hegemony. As a political and cultural construction, the 'general interest' aligns the needs of subalterns (for jobs, subsistence, recreation, etc.) with the particular, long-term interests of capital, in what Jessop (2024:269) terms a hegemonic project which 'can resolve the abstract problem of conflicts between particular interests and the general interest.' As Gramsci (1971:161) noted, in this resolution, the 'decisive nucleus' of capital accumulation must be protected and reproduced. Within that important constraint, a hegemonic project that wins popular support typically involves 'the sacrifice of certain short-term interests of the hegemonic class' and 'a flow of material concessions for other social forces mobilized behind the project' – all of which is conditioned and limited by the accumulation process (ibid.).

Importantly, the translation of particular interests into a general interest entails more than material concessions; it extends into what Gramsci called the ethico-political – the realm of values, social visions and identity. This does not mean that a rosy consensus is reached among all. The construction of a general interest organizes capital's allies into a bloc supporting the regime, yet 'those particular interests that are inconsistent with the project are deemed immoral and/or irrational and, insofar as they are still pursued by groups outside the consensus, they are also liable to sanction' (Jessop 2024:269). In short, 'hegemonic social blocs express their interests as the interests of society as a whole' (Ekers and Prudham 2018:24).

Crucially, to reiterate a point above, a viable hegemonic project must fit with the requirements of capital accumulation. A hegemonic project, thus, is a complex of policy frameworks and political visions of the good life, extending to the activities of a host of non-state organizations in civil society (from Rotary Clubs through churches, to many social-media platforms), and to common-sense understanding and desires of the subalterns aligned with the project. When such alignment is achieved – satisfying the needs of capital for self-expansion and incorporating (many of) the masses – one can speak of an 'historical bloc' that secures stable support for a capitalist regime. Within the bloc, translation of narrow, particular interests into broader ethico-political ones 'not only helps to co-constitute economic structures but also provides them with their rationale and legitimacy' (Sum and Jessop 2013:199).

In the next chapter I will trace the successful construction, under post-World War 2 Pax Americana, of the historical bloc supporting Fossil Capitalism in the advanced capitalist North, organized around what Brant and Wissen term 'the imperial mode of living.' The pathway towards this was forged in the preceding eras of global capitalist expansion, as hegemonic projects featuring liberal promises of freedom and Promethean celebrations of progress attained persuasive power within world capitalism's core, to be exported to the global South as 'modernization' after World War 2. In these visions, carbon extraction and fossil-capital accumulation became common sense as heavy industry, new modes of transport and the expanding commodification of life transformed both objective and subjective realities. As

Fossil Capitalism developed, the massive accumulation of fixed capital – coal mines, railways and coke ovens, later oil rigs and tankers, pipelines, refineries, gas stations, etc., and other durable aspects of accumulation – transformed not only landscapes. These enduring parts of the built environment gave form to the 'ideological pillars of legitimacy, including notions of freedom, modernity, progress, and the like, notions that in turn are foundational to stitching together hegemonic social orderings via the expression of the interests of leading groups and classes as the ideals, values, and beliefs of society more generally' (Ekers and Prudham 2018:14).

To summarize, when we speak of hegemony with reference to Fossil Capitalism we refer to the complexes of practices and relations, centred upon state apparatuses but extending integrally into civil society, that have blended consent with coercion in managing Fossil Capitalism within multi-scalar historical blocs (as in Carroll 2021a:481ff). Hegemony's multi-scalarity is what cements the organization of consent, creating a durable order in which opposition is marginalized and diffused. As we shall see in Chapter 2, from suburban lifestyles in the global North entirely predicated on fossil power to transnational organizations and networks that legitimate fossil power as integral to modern life, oil in particular has been elevated to the status of our 'lifeblood' (Huber 2013).

The Rise of Carbon Democracy

Fossil Capitalism has thus developed, at different scales, within a matrix of power relations combining economic power (capital accumulation) and political/cultural power (hegemonic projects), in a changing context of imperialist world order. Our next two chapters will explore these developments in depth. At this point, to underline the interplay of accumulation, imperialism and hegemony in the making of Fossil Capitalism, I turn to Timothy Mitchell's (2011) influential study of carbon democracy.

'Democracy' is a term at the centre of modern hegemonic discourse. But 'like energy from fossil fuels, democratic politics is a recent phenomenon. The development of the two kinds of power has been interwoven from the start' (Mitchell 2011:8). Mitchell sets himself the task of unravelling how these two forms of power were 'co-assembled', beginning in the late 19th century with coal and the emergence of mass politics in capitalism's north Atlantic heartland (ibid.). This was the period British historian Eric Hobsbawm called both 'the age of democratization' *and* 'the age of empire' – and for good reason: the two processes were indeed interwoven. The same period, extending into early 20th century, was the era of the Second Industrial Revolution (Hull 1999), as technological innovations in steel production, chemicals, transportation and communication and electrification transformed production and quickened the pace of urbanization in the capitalist North.

By the 1870s, coal-based industrial capitalism had been consolidated in Europe and North America, along with systems of representative 'democracy' so exclusionary as to invert the meaning of the term. As Mitchell explains,

> The relationship between coal, industrialisation and colonisation provides a first set of connections between fossil fuels and democracy. Forms of representative central government had developed in parts of Europe and its settler colonies in the eighteenth and nineteenth centuries. The advocates of representative government had seen it not as a first step towards democracy but as an oligarchic alternative to it, in which the power of government was reserved to those whose ownership of property (the control of land, but also of women, servants and slaves) gave them power over the point of passage for the revenues on which government depended, and qualified them to be concerned with public matters.
>
> *(2011:17)*

While the leading capitalist states embraced a discourse of democracy, the historical bloc that these regimes of oligarchy-masquerading-as-democracy organized was extremely limited. It excluded the majority from participation in public life, while enforcing the market logic that consigned growing numbers to chronic poverty (Acemoglu and Robinson 2000). Meanwhile, the second industrial revolution brought great numbers of property-less proletarians together – into factories and mines and into cramped urban quarters. A strategically significant site in this regard was the many coalmines wherein fuel powering the industrial revolution was extracted. As Mitchell notes, 'great quantities of energy now flowed along very narrow channels' (2009:403), with many workers concentrated at the junctions of those channels, affording them 'at certain moments, a new kind of political power. The power derived not just from the organizations they formed, the ideas they began to share or the political alliances they built, but from the extraordinary concentrations of carbon energy whose flow they could now slow, disrupt or cut off' (ibid.).

The disruptive power of a strategically positioned fraction of the working class was fundamental in demanding an extension of democratic participation to proletarians, along with reforms 'whose gradual implementation radically reduced the precariousness of life in industrial societies' (Mitchell 2011:27). Coal miners led the struggle, 'contesting work regimes and the private powers of employers in the labour activism and political mobilisation of the 1880s and onward' (Mitchell 2011:19).

So it was that, within the core of the world system, the dialectic of class struggle, built into Fossil Capitalism, shaped a political force that contested the limited hegemonic project of oligarchic democracy. Coal mining brought workers together at choke-points in the fossil-capital commodity chain, 'enhancing their power and enabling the working class in the Global North to demand concessions that led to "carbon democracy"' (Carroll 2021b:6). However, as Mitchell also observes, the framework for democracy that was solidified through these struggles, continuing well into the 20th century, still suited the interests of the dominant class. This form of democracy is 'a mode of governing populations that employs popular consent as a means of limiting claims for greater equality and justice by dividing up the common world' (Mitchell 2011:9). Carbon democracy broadened Fossil Capitalism's historical bloc, in a way that reinforced ruling class hegemony, keeping certain fields (markets, private property) under nondemocratic control by propertied employers and investors. Democratic struggles became 'a battle over the distribution of issues' (ibid.): 'border conflicts' (Fraser 2022) over which claims can be taken up democratically and which fall within the purview of capital.

Mitchell's analysis of carbon democracy is important not only for its depiction of the close relationship between the changing shape of hegemony in the North and the consolidation of Fossil Capitalism, but for its insights on the deep relationship between these developments and the establishment of modern imperialism. As he put it, 'the rise of coal produced democracy at some sites and colonial domination at others' (2011:18). As I discussed earlier (see Table 1.1), the final decades of the 19th century and first decades of the 20th marked the emergence of monopoly capitalism and modern imperialism, concurrently with a transition in which petroleum began to gain prominence as an energy source. As a reminder of the continuing imbrication of fossil capital and imperialism, Mitchell's book was written in the context of the US war on Iraq (2003–11), a classic example of oil imperialism.

More recently, in an interview, Mitchell summarized the thesis of *Carbon Democracy* in a sentence: 'coal made possible the emergence of mass democracy, and oil set its limits' (Szeman and Wellum 2023:354).

Two such limits stand out. On the one hand, the fossil-powered world of industrial capitalism 'was also a colonising world' (Mitchell 2011:84) – and colonialism is obviously antithetical to democracy. Even as coal enabled the concentration of industrial production at sites close to the coal mines of Europe and North America, the need for materials only available elsewhere 'encouraged the expansion of mining, plantations and colonial settlement across wide areas of the non-European world, along with railways, banking firms, investment capital and imperial armies' (ibid.). The rush to control resources from the majority world eventuated in a division of the majority world among a few colonizing powers, most notably in the obscene 'scramble for Africa' beginning in the late 19th century (Campbell 2015). The rather thin 'carbon democracy' being rolled out in what was becoming the global North was entirely denied to the colonized peoples comprising most of humanity.

On the other hand, the rise of petroleum as a fossil fuel (at first limited to kerosene lighting, but gradually becoming more generally deployed) brought into play different labour processes that weakened workers' collective capacities to press for democratization. In mining coal, workers experience considerable autonomy in a labour process that mostly occurs underground. In extracting oil, workers remain above ground, under supervision of managers. Moreover, the production and refining of petroleum is highly capital-intensive, requiring far fewer workers, and, since oil is a liquid, its transport also uses less human labour. All these differences weakened the disruptive power of collective labour, as Fossil Capitalism's predominant energy source began to shift. But a major point of weakness, strongly reinforcing the anti-democratic thrust of imperialism, was capital's divide-and-rule strategy in organizing extractive capitalism in west Asia (and elsewhere).

> The ability to weaken the labour force by dividing it into separate racial groups, with managers, skilled workers and unskilled workers housed and treated separately, reflected the different distribution of oil production across the world compared to coal, and its development after rather than before the rise of modern industry. Oil production often grew rapidly, in regions remote from large populations, to serve distant users in places already industrialised with coal – a fact that encouraged the producers to import workers from different places and then perpetuate the forms of ethnic division.
>
> *(Mitchell 2011:35–6)*

Meanwhile, in the battle for hearts and minds that is central in hegemonic struggle, the major oil companies 'were learning to portray their needs as furthering the imperial interests of the state, and thus contributing to the well-being of the nation' (Mitchell 2011:54). As Engels, reflecting on the condition of the working class in England, observed already in 1892:

> The truth is this: during the period of England's industrial monopoly the English working-class have, to a certain extent, shared in the benefits of the monopoly. These benefits were very unequally parcelled out amongst them; the privileged minority pocketed most, but even the great mass had, at least, a temporary share now and then.
>
> *(Engels 1892)*

Many of those benefits issued from ecological unequal exchange between the imperial centre and the capitalist periphery, where the destructive impacts of extractive accumulation were primarily experienced (Gago and Mezzadra 2017; Nygren et al. 2022). The imperialist structure of accumulation thus furnished a basis for a hegemonic historical bloc that came to include the higher strata of the northern proletariat, who experienced not only 'carbon democracy' but some measure of material affluence – both of which contributed to their ideological integration into capitalism as a way of life. Forged in the early 20th century, these basic features of capitalist hegemony in the global North continue to resonate a century later.

In the course of the first decades of the 20th century, 'a handful of industrialised states in the global north had brought much of the world under the control of imperial government' (Mitchell 2011:68). Mitchell goes on to note that the resources that enabled modern imperialism also gave groups of organized workers in the advanced capitalist states 'an unusual power to make successful political claims' (ibid.). In the global South, however, colonial domination became further consolidated with the 20th-century transition from coal to oil, as seven oil majors came to control supply, engendering 'a geopolitics of domination in which the US figured prominently' (Williams 2018:237). Tied together through fossil-capital accumulation, carbon democracy in the North and ecological imperialism in the South came to form a transnational system of exploitation and expropriation.

Conclusion

The historical-materialist scholarship I have taken up in this chapter demonstrates how, through a trifecta of interlinked, power-laden processes – namely, capital accumulation, colonialism/imperialism and hegemonic struggle – Fossil Capitalism emerged in the 19th century and advanced further in the 20th century. It is through these interlinked processes that a world order was consolidated and protected – a way of life that is ecocidal. As capitalist accumulation, powered by fossilized energy, has scaled up to global level it has pressed against ecological limits and has begun to eradicate the very bases for living systems – the climate crisis being the most fateful aspect of this ecocide (Whyte 2020).

Andreas Malm, drawing on Wisner et al.'s (2005) conceptual framework for understanding disasters, has presented a useful dialectical model of climate disaster. The model, shown as Figure 1.1, depicts a convergence of ecological hazards from the progression of global heating and social hazards from the progression of human vulnerabilities, all of them shaped by capitalist development.

As this diagram suggests, and as the chapters that follow will show, we are truly a crossroads. In 1930, Antonio Gramsci remarked, in the context of an earlier organic crisis, at a different crossroads, 'the old is dying and the new cannot be born; in this interregnum a great variety of morbid symptoms appear' (1971:276). The morbid symptoms today are far more horrific than Gramsci could have imagined. Will Fossil Capitalism continue along its ecocidal path? Within the global order, will ruling classes find a way towards a fifth era for imperialism (with all the human and ecological nastiness that implies), politically stabilized somehow through constructing a new historical

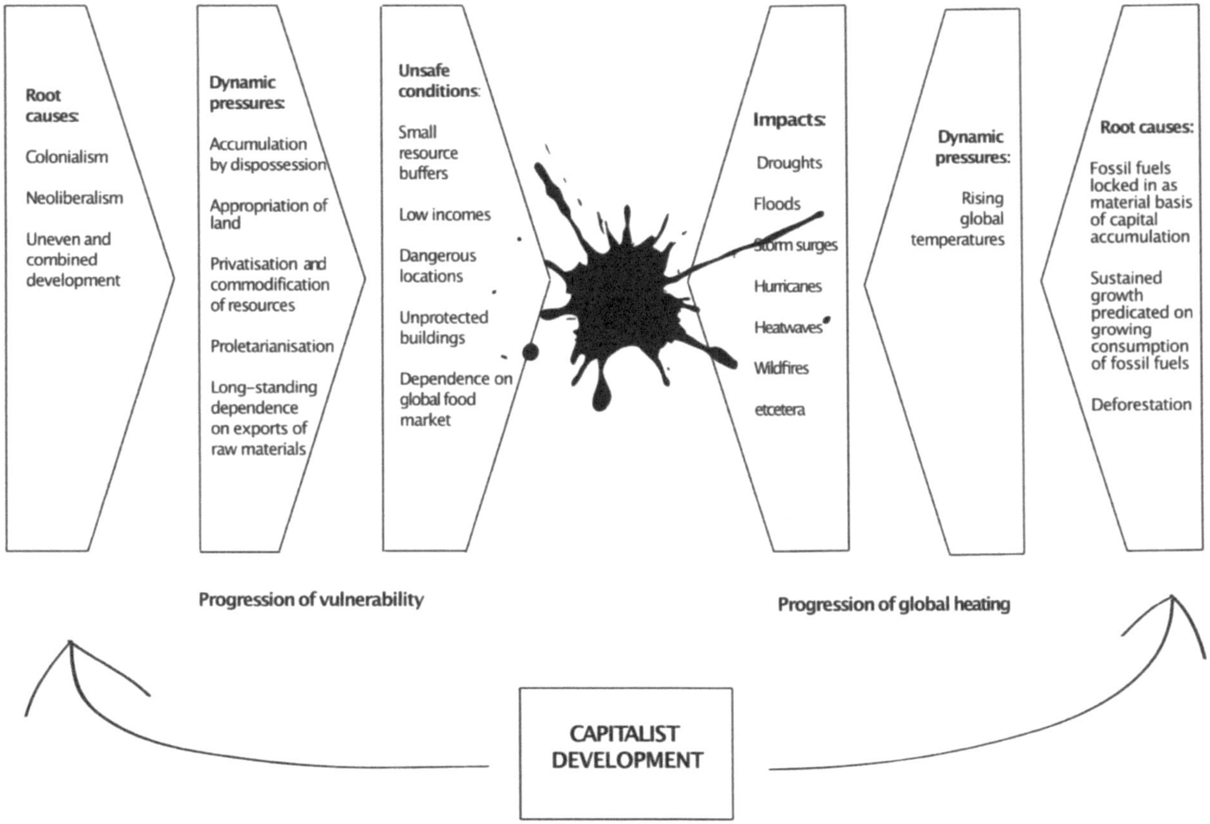

FIGURE 1.1 A Dialectical Model of Climatic Disaster.

Source: Malm (2020:70).

bloc? Or will anti-imperialist and other progressive currents form a collective will strong enough to enable a transcendence of capitalism and imperialism? Having developed the analytical tools for addressing these issues, let us turn to the post-World War 2 era, a period in which carbon democracy became elaborated into the imperial mode of living that many people in the global North now take for granted (or aspire to), and that many in the South dream of.

Notes

1 There is irony in this claim. After all, the world in which we find ourselves is permeated by obvious inequities, injustices and ecological calamities. Good social science requires the investigation of how modernity's maladies have come about, what practices stabilize and reproduce them and how they can be undone and transcended. Historical materialism directly addresses these issues, as no other social-scientific perspective does. As I have put things elsewhere, '*in a socially unjust world, knowledge of the social that does not challenge injustice is likely to play a role in reproducing it*' (Carroll 2004: 3). Intellectuals who dismiss HM as unscientific play this role.

2 Since labour power is literally embodied in human beings, its value is determined not only by bare subsistence needs but by class struggle over wages rates (Marx 1976a).

3 As Sadi Carnot, writing at the dawn of Fossil Capitalism put matters, coal 'has the invaluable advantage of being employable at any time and in any place, and of never suffering an interruption in its work' (Carnot 1824:2, quoted in Fressoz and Bonneuil 2017:55).

4 O'Rourke (2018) lists 64 such covert operations involving the CIA, from its creation in 1947 to 1989. For context, in 1960 there were 108 sovereign states in the world (Butcher and Griffiths 2020).

5 See Gramsci (1975). A serviceable English translation of selections from the Notebooks was published in 1971 (Gramsci 1971). The complete Notebooks have yet to be translated into English (but see Gramsci 2011).

References

Acemoglu, Daron, and James A. Robinson. 2000. "Why Did the West Extend the Franchise? Democracy, Inequality, and Growth in Historical Perspective." *Quarterly Journal of Economics* 115(4):1167–99.

Altvater, Elmer. 2006. "The Social and Natural Environment of Fossil Capitalism." *Socialist Register* 43:37–59.

Angus, Ian. 2016. *Facing the Anthropocene*. New York: Monthly Review Press.

Arsel, Murat, and Lorenzo Pellegrini. 2022. "Global Extractive Imperative: From Local Resistance to Unburnable Fuels." *International Development Planning Review* 44(1):1–12. doi: 10.3828/idpr.2021.25

Beeson, Mark, and André Broome. 2010. "Hegemonic Instability and East Asia: Contradictions, Crises and US Power." *Globalizations* 7(4):507–23. doi: 10.1080/14747731003798500

Block, Fred. 1977. "The Ruling Class Does Not Rule: Notes on the Marxist Theory of the State." *Socialist Revolution* 33: 6–28.

Bonanno, Alessandro. 2000. "The Crisis of Representation: The Limits of Liberal Democracy in the Global Era." *Journal of Rural Studies* 16(3):305–23. doi: 10.1016/S0743-0167(99)00064-9

Bond, Patrick, and Federico Fuentes. 2024. "BRICS: Anti-Imperialist or Sub-Imperial Allies?" *Green Left* 1397 January 3. Retrieved August 27, 2024 (https://www.greenleft.org.au/content/brics-anti-imperialist-or-sub-imperial-allies).

Bowles, Paul, and Henry Veltmeyer. 2017. "Extractivisms: Concepts, Local Impacts and Spill-Effects." Pp. 37–52 in *Alternative Development Models: Latin American Results and Prospects*, edited by R. Munck and H. Veltmeyer. London: Routledge.

Brancaccio, Emiliano. 2023. "Why Don't Capitalists Want to 'Go Green' Anymore?" *The Bullet*. Retrieved February 29, 2024 (https://socialistproject.ca/2023/10/why-dont-capitalists-want-to-go-green-anymore/).

Brand, Ulrich, and Markus Wissen. 2024. "Fordism, Post-Fordism and the Imperial Mode of Living." Pp. 279–97 in *The Elgar Companion to Antonio Gramsci*, edited by W. K. Carroll. Cheltenham, UK; Northampton, MA: Edward Elgar Publishing.

Brewer, Anthony. 1980. *Marxist Theories of Imperialism: A Critical Survey*. 1st ed. London: Routledge.

Buddharaksa, Watcharabon. 2018. "Big Society, Free Economy, and Strong State: Bonefeld's Open Marxism and the Critique of Political Economy." *Journal of Social Sciences* 48(1):59–74.

Buller, Adrienne. 2022. *The Value of a Whale: On the Illusions of Green Capitalism*. Manchester: Manchester University Press.

Burawoy, Michael. 1990. "Marxism as Science: Historical Challenges and Theoretical Growth." *American Sociological Review* 55(6):775–93.

Butcher, Charles R., and Ryan D. Griffiths. 2020. "States and Their International Relations since 1816: Introducing Version 2 of the International System(s) Dataset (ISD)." *International Interactions* 46(2):291–308. doi: 10.1080/03050629.2020.1707199

Campbell, Horace. 2015. "Imperialism and Anti-Imperialism in Africa." *Monthly Review* 67(3):98. doi: 10.14452/MR-067-03-2015-07_7

Carnot, Sadi. 1824. *Réflexions Sur La Puissance Motrice Du Feu*. Paris: Bachelier.

Carroll, William K., ed. 2004. *Critical Strategies for Social Research*. Toronto: Canadian Scholars' Press.

Carroll, William K. 2020. "Fossil Capital, Imperialism and the Global Corporate Elite." Pp. 30–57 in *BRICS and the New American Imperialism*, edited by V. Satgar. Johannesburg: Wits University Press.

Carroll, William K., ed. 2024. *The Elgar Companion to Antonio Gramsci*. Cheltenham, UK; Northampton, MA: Edward Elgar Publishing.

Carroll, William K. 2021a. "Conclusion: Prospects for Energy Democracy in the Face of Passive Revolution." Pp. 479–504 in *Regime of Obstruction*, edited by W. K. Carroll. Edmonton: AU Press.

Carroll, William K. 2021b. "Introduction." Pp 3–31 in *Regime of Obstruction*, edited by W. K. Carroll. Edmonton: AU Press.

Carroll, William K., and J. P. Sapinski. 2018. *Organizing the 1%: How Corporate Power Works*. Black Point, Nova Scotia; Winnipeg, Manitoba: Fernwood Publishing.

Carson, Rachel. 2002. *Silent Spring*. 50th anniversary ed. Boston, MA: Mariner Books.

Clark, Brett, and Richard York. 2005. "Carbon Metabolism: Global Capitalism, Climate Change, and the Biospheric Rift." *Theory and Society* 34(4):391–428.

Cox, Robert W. 1987. *Production, Power, and World Order: Social Forces in the Making of History*. New York: Columbia University Press.

Derber, Charles. 2010. *Greed to Green*. London: Paradigm Publishers.

Dunlap, Alexander. n.d. "The Structures of Conquest: Debating Extractivism(s), Infrastructures and Environmental Justice for Advancing Post-Development Pathways." Pp. 52–76 in *The Afterlives of Extraction: Alternatives and Sustainable Futures*, edited by F. Calvão, M. Archer, and A. Benya. Leiden: Brill-Nijhoff.

Ekers, Michael, and Scott Prudham. 2018. "The Socioecological Fix: Fixed Capital, Metabolism, and Hegemony." *Annals of the American Association of Geographers* 108(1):17–34. doi: 10.1080/24694452.2017.1309963

Energy Institute. 2023. *Statistical Review of World Energy*. London: Energy Institute.

Engels, Frederick. 1892. "Preface to the English Edition." In Frederick Engels, *The Condition of the Working Class in England. Marx/Engels Internet Archive*. Accessed August 28, 2023 (https://www.marxists.org/archive/marx/works/1892/01/11.htm).

Foster, John Bellamy. 1999. "Marx's Theory of Metabolic Rift: Classical Foundations for Environmental Sociology." *American Journal of Sociology* 105(2):366–405. doi: 10.1086/210315

Foster, John Bellamy. 2022. "The Return of the Dialectics of Nature: The Struggle for Freedom as Necessity." *Historical Materialism* 30(2):3–28.

Foster, John Bellamy, and Batuhan Sarican. 2023. "'Monthly Review' and the Environment." *Monthly Review* 75(6):22.

Frame, Mariko L. 2022. "Ecological Imperialism: A World-Systems Approach." *The American Journal of Economics and Sociology* 81(3):503–34. doi: 10.1111/ajes.12472

Fraser, Nancy. 2022. *Cannibal Capitalism: How Our System Is Devouring Democracy, Care, and the Planet – and What We Can Do about It*. London, New York: Verso.

Fressoz, Jean-Baptiste, and Christophe Bonneuil. 2017. "Growth Unlimited: The Idea of Infinite Growth from Fossil Capitalism to Green Capitalism." Pp. 52–68 in *History of the Future of Economic Growth: Historical Roots of Current Debates on Sustainable Degrowth*, edited by I. Borowy and M. Schmelzer. London; New York, NY: Routledge, Taylor & Francis Group.

Gago, Verónica, and Sandro Mezzadra. 2017. "A Critique of the Extractive Operations of Capital: Toward an Expanded Concept of Extractivism." *Rethinking Marxism* 29(4):574–91. doi: 10.1080/08935696.2017.1417087

Ghosh, Jayati, Shouvik Chakraborty, and Debamanyu Das. 2022. "Climate Imperialism in the Twenty-First Century." *Monthly Review* 70–85. doi: 10.14452/MR-074-03-2022-07_4

Graham, Nicolas. 2021. *Forces of Production, Climate Change and Canadian Fossil Capitalism*. Leiden: Brill.

Graham, Nicolas, and William K. Carroll. 2022. "Climate Breakdown: From Fossil Capitalism to Climate Capitalism (and Beyond?)." *Socialist Register 2023* 59:25–46.

Gramsci, Antonio. 1971. *Selections from the Prison Notebooks*. edited by Q. Hoare and G. Nowell-Smith. New York: Lawrence & Wishart.

Gramsci, Antonio. 1975. *Quaderni Del Carcere (Prison Notebooks)*. Critical Edition, edited by V. Gerratana. Turin: Einaudi.

Gramsci, Antonio. 2011. *Prison Notebooks*. Vols. 1–3. edited by J. A. Buttigieg. New York: Columbia University Press.

Harvey, David. 2003. *The New Imperialism*. Oxford, UK: Oxford University Press.

Hersh, Seymour. 2023. "How America Took Out The Nord Stream Pipeline." *Seymour Hersh*. Retrieved April 22, 2024 (https://seymourhersh.substack.com/p/how-america-took-out-the-nord-stream).
Hoare, Q., and G. N. Smith. 1971. "Introduction: The Intellectuals." Pp. 3–4 in *Selections from the Prison Notebooks of Antonio Gramsci*, edited by Q. Hoare and G. N. Smith. New York: International Publishers.
Hobsbawm, Eric. 1989. *The Age of Empire: 1875–1914*. New York: Vintage Books.
Hope, Wayne. 2018. "Epochality, Global Capitalism and Ecology." *TripleC: Communication, Capitalism & Critique* 16(2):562–76. doi: 10.31269/triplec.v16i2.1002
Huber, Matthew T. 2013. *Lifeblood: Oil, Freedom, and the Forces of Capital*. Minneapolis: University of Minnesota Press.
Hull, James. 1999. "The Second Industrial Revolution: The History of a Concept." *Rivista Internazionale Di Storia Della Storiografia* 36:81–90.
Jessop, Bob. 2024. "Hegemonic Projects and Cultural Political Economy." Pp. 261–78 in *The Elgar Companion to Antonio Gramsci*, edited by W. K. Carroll. Cheltenham, UK; Northampton, MA: Edward Elgar Publishing.
Kenneth, A. Gould, David N. Pellow and Allan Schnaiberg. 2004. "Interrogating the Treadmill of Production: Everything You Wanted to Know about the Treadmill but Were Afraid to Ask." *Organization & Environment* 17(3):296–316.
Larrain, Jorge. 1986. *A Reconstruction of Historical Materialism*. London: Allen and Unwin.
Lebowitz, Michael A. 2020. *Between Capitalism and Community*. New York: Monthly Review Press.
Lewontin, Richard, and Richard Levins. 2007. *Biology under the Influence: Dialectical Essays on the Coevolution of Nature and Society*. New York: NYU Press.
Liebig, Justus von. 1859. *Letters on Modern Agriculture*. London: Walton & Maberly.
Luke, Timothy W. 2006. "The System of Sustainable Degradation." *Capitalism Nature Socialism* 17(1):99–112. doi: 10.1080/10455750500505556
Malm, Andreas. 2016. *Fossil Capital: The Rise of Steam Power and the Roots of Global Warming*. London: Verso.
Malm, Andreas. 2020. *Corona, Climate, Chronic Emergency: War Communism in the Twenty-First Century*. London: Verso.
Mandel, Ernest. 1976. "Introduction." Pp. 11–86 in Ben Fowke (translator) Karl Marx, *Capital*. Vol. 1. London: Penguin Books.
Marx, Karl. 1843. "Letter from Marx to Arnold Ruge." *Marx/Engels Internet Archive*. Retrieved August 27, 2023 (https://www.marxists.org/archive/marx/works/1843/letters/43_09-alt.htm).
Marx, Karl. 1852. *The Eighteenth Brumaire of Louis Bonaparte*. Moscow: Progress Publishers.
Marx, Karl. 1976a. *A Contribution to the Critique of Political Economy*. New York: International Publishers.
Marx, Karl. 1976b. *Capital*. Vol. 1. London: Penguin Books.
Marx, Karl. 2002. "Theses on Feuerbach." *Marx/Engels Internet Archive*. Retrieved May 14, 2024 (https://www.marxists.org/archive/marx/works/1845/theses/).
Marx, Karl. 1959. *Economic and Philosophical Manuscripts of 1844*. Moscow: Progress Publishers. Retrieved August 27, 2024 (https://www.marxists.org/archive/marx/works/download/pdf/Economic-Philosophic-Manuscripts-1844.pdf).

Marx, Karl, and Friedrich Engels. 1965. *The German Ideology*. London: Lawrence & Wishart.

Marx, Karl, and Friedrich Engels. 2019. "Manifesto of the Communist Party." *Marx/Engels Internet Archive*. Retrieved May 14, 2024 (https://www.marxists.org/archive/marx/works/1848/communist-manifesto/).

McNally, David. 2006. *Another World Is Possible: Globalization & Anti-Capitalism*. 2nd ed., rev. expanded ed. Winnipeg, Manitoba: Arbeiter Ring Publ.

McNeill, John Robert. 2000. *Something New Under the Sun: An Environmental History of the 20th Century*. New York: W.W. Norton & Co.

McNeill, John Robert, and Peter Engelke. 2014. *The Great Acceleration: An Environmental History of the Anthropocene since 1945*. Cambridge, MA: The Belknap Press of Harvard University Press.

Miliband, Ralph. 1983. "State Power and Class Interests." *New Left Review* 138:57–68.

Mitchell, Timothy. 2009. "Carbon Democracy." *Economy and Society* 38(3):399–432. doi: 10.1080/03085140903020598

Mitchell, Timothy. 2011. *Carbon Democracy*. London: Verso.

Moore, Jason. 2016. "The Rise of Cheap Nature." Pp. 78–115 in *Anthropocene or Capitalocene: Nature, History and the Crisis of Capitalism*, edited by J. Moore. Oakland, CA: PM Press.

Muhr, Thomas. 2024. "Gramsci and Hegemonic Struggle in a Globalized World." Pp. 406–26 in *The Elgar Companion to Antonio Gramsci*, edited by W. K. Carroll. Cheltenham, UK; Northampton, MA: Edward Elgar Publishing.

Nygren, Anja, Markus Kröger, and Barry Gills. 2022. "Global Extractivisms and Transformative Alternatives." *The Journal of Peasant Studies* 49(4):734–59. doi: 10.1080/03066150.2022.2069495

O'Connor, James. 1988. "Capitalism, Nature, Socialism: A Theoretical Introduction." *Capitalism Nature Socialism* 1(1):11–38. doi: 10.1080/10455758809358356

O'Rourke, Lindsey A. 2018. *Covert Regime Change: America's Secret Cold War*. Ithaca, New York: Cornell University Press.

Pijl, Kees van der. 1998. *Transnational Classes and International Relations*. London, New York, NY: Routledge.

Pineault, Éric. 2023. *A Social Ecology of Capital*. London: Pluto Press.

Robinson, William I. 2015. "The Transnational State and the BRICS: A Global Capitalism Perspective." *Third World Quarterly* 36(1):1–21.

Rudy, Alan P. 2019. "On Misunderstanding the Second Contradiction Thesis." *Capitalism Nature Socialism* 30(4):17–35. doi: 10.1080/10455752.2019.1652663

Schnaiberg, Allan. 1980. *The Environment: From Surplus to Scarcity*. New York, NY: Oxford Univ. Press.

Smil, Vaclav. 2010. *Energy Transitions: History, Requirements, Prospects*. Santa Barbara, CA: Praeger.

Spence, Martin. 2000. "Capital Against Nature: James O'Connor's Theory of the Second Contradiction of Capitalism." *Capital & Class* 24(3):81–110. doi: 10.1177/030981680007200105

Sum, Ngai-Ling, and Bob Jessop. 2013. *Towards a Cultural Political Economy: Putting Culture in Its Place in Political Economy*. Northampton, MA: Elgar.

Szeman, Imre, and Caleb Wellum. 2023. "Carbon Democracy at Ten: An Interview with Timothy Mitchell." *Cultural Studies* 37(3):351–69. doi: 10.1080/09502386.2022.2056221

Tabak, Faruk. 2015. *Allies As Rivals*. London: Routledge.
Wargan, Paweł. 2023. "NATO and the Long War on the Third World." *Monthly Review* 16–32. doi: 10.14452/MR-074-08-2023-01_2
Weeks, John. 1981. "The Differences between Materialist Theory and Dependency Theory and Why They Matter." *Latin American Perspectives* 8(3/4):118–23.
Whyte, David. 2020. *Ecocide: Kill the Corporation before It Kills Us*. Manchester: Manchester University Press.
Williams, Michelle. 2018. "Energy, Labour and Democracy in South Africa." Pp. 231–51 in *The Climate Crisis: South African and Global Democratic Eco-socialist Alternatives*, edited by V. Satgar. Johannesburg: Wits University Press.
Wisner, Ben, Piers Blaikie, Terry Cannon, and Ian Davis. 2005. *At Risk: Natural Hazards, People's Vulnerability and Disasters*. 2nd ed. London: Routledge.
Wood, Ellen Meiksins. 1981. "The Separation of the Economic and the Political in Capitalism." *New Left Review* 127: 66–95.
World Energy Council. 2016. *World Energy Resources 2016*. London: World Energy Council.

2

FORDISM, CONSUMER CAPITALISM AND THE GREAT ACCELERATION

Capitalism, as James O'Connor (1987) emphasized, is a crisis-prone, crisis-ridden and crisis-dependent form of economy. Crises are 'the cauldrons in which capital qualitatively restructures itself for economic, social, and political renewal and further accumulation' (1987:93–4). The rhythm of accumulation moves in waves, both the short waves of the standard business cycle (booms and busts roughly every decade) and the 'long waves' that involve deeper depression and within that, more extensive capital restructuring (Mandel 1975). The Great Depression of the 1930s was the downside of a long wave of early-20th Century accumulation in which monopoly capitalism and classic imperialism emerged out of the Long Depression of 1873–96 (Kaya 2022; Hobsbawm 1989). As the 'Dirty Thirties' closed out, Depression morphed into world war, stoking militarized accumulation (particularly in the United States) but ultimately destroying enormous amounts of fixed capital (and human populations) in Europe and east Asia. After six years of world war, in the later 1940s global capitalism was reconstructed amid new conditions that favoured highly profitable accumulation. A classic illustration of capitalism's crisis-dependency, in destroying capital that would otherwise had claimed slices of surplus value, thereby depressing profits, Depression and war created vast opportunities for new investment that would incorporate the newest technologies (Harvey 2006:200–3, 444).

The post-World War 2 'golden years' of unprecedented economic boom, reaching into the 1970s, reconfigured important aspects of accumulation, imperialism and hegemony. The production and consumption of fossil energy was a vital force in all these transformations, as was the widening of the gap between affluent North and imperialized South, which had been expanding since the industrial revolution.[1] As our earlier discussion of the Great

DOI: 10.4324/9781003412922-4

Acceleration implied, for Fossil Capitalism, this was a pivotal era, bearing close scrutiny. If carbon democracy arose during the era of classic imperialism (see Table 1.1), it was not until the aftermath of World War 2 that the marriage of Fossil Capitalism and liberal democracy was consummated and normatively universalized across the global North. The defeated Axis powers – the fascist regimes of Germany, Italy, Japan and lesser allies – had based their alliance on strident anti-communism, but they were also opponents of liberal democracy. The Soviet Union – the Axis's main target – made the largest contribution to the victory over fascism but also took by far the greatest hit, losing more than 20 million citizens and much of its industrial base. Indeed, by the close of the war, in September 1945, much of Europe and Japan lay in ruins. Conversely, the United States, which with the exception of the Pearl Harbor bombing (December 7, 1941) emerged unscathed from actual fighting on its territory, benefitted enormously from the war. Its industries, already relatively advanced by the 1930s, surged forward during the wartime economy, in both scale and technology.

In the immediate post-war years, the United States, the last major imperialist power standing after World War 2's barbarism, cemented its international hegemony through several means. The goal was 'to create conditions favorable to the expansion of its economy, and to the growth of the Western camp in general' (Fressoz and Bonneuil 2017:59–60).

> It was in this context that a new international economic regime was established, based on free trade and growth: the Bretton Woods agreements of 1944 established the dollar as world currency, the General Agreement on Tariffs and Trade liberalized trade in 1947, coupled with the Marshall Plan and the Point Four Program of the Truman Administration, which created the concept of "development aid." This world order made it possible to find outlets for the United States' gigantic industrial and agribusiness production and ensured full employment and social pacification after the great strikes of 1946. It also aimed at the social stabilization of the Western world by drawing it into growth.
>
> *(Fressoz and Bonneuil 2017)*

New international institutions for managing capitalism emerged under US leadership, including the World Bank and the International Monetary Fund. The North Atlantic Treaty Organization (1949) and the United States' continuing military occupation (continuing actually to this day) of Germany and Japan enabled the American hegemon to project its military power globally, on a permanent basis.[2] Concurrently, trade liberalization broke up the European colonial empires, opening access for the rapidly growing US-based TNCs to the fossil and other resources of former colonies.

The early post-war years witnessed construction of new energy networks that replaced coal with oil, and a deepening of oil imperialism in west Asia (Mitchell 2009:409–13). Oil became Fossil Capitalism's 'lifeblood' (Huber 2013) in the context of profound political-economic and ideological transformations that included a new, abstract conception of 'the economy.' Measured quantitatively as Gross National Product (GNP), it became 'an object that could grow without limit', as the falling price of oil and its relative abundance removed any need to attend to the depletion of reserves (Mitchell 2009: 418).

The Hegemony of Endless Growth and Cold War Liberalism

Indeed, 'growth' became an obsession for state managers concerned with the possibility of a post-war depression. In national accounting, GNP (later GDP) naturalized the notion of the economy as 'a closed circuit, a circular flow of value between production and consumption cut off from its natural moorings.' Concomitantly, 'by measuring it with just one figure, national accounting reified the economy and made it possible to erect it into an entity separate from society, politics, and nature' (Fressoz and Bonneuil 2017:58–9). This reification, central to late capitalist hegemony, continues to this day, as any perusal of business news discourse, obsessed with prospects for economic growth, will show.

Within the United States, the Paley Commission (1951–52), tasked with finding a solution to the increasingly evident problem of resource scarcity, issued an influential report which elevated the idea of 'the economy' as 'a discrete entity whose operation was defined by theoretically limitless growth.' This was 'a vital step in formally establishing economic growth, hitherto a subjective value, as the defining priority of government' (Buller 2022:53). Moreover,

> through a technical revision in how resource scarcity was calculated, the report both intellectually validated and politically prioritised not only growth but also the acceleration of globalisation and American imperialism in the extraction and provision of resources, particularly overseas fossil fuel development.
>
> *(Buller 2022)*

In launching the Great Acceleration of fossil fuel consumption and expropriation of nature that came to define the latter half of the twentieth century, the political embrace of endless growth at the core of the post-war hegemonic project set the table for climate crisis. As capital accumulation exploded, the period of the Great Acceleration witnessed

> an astonishing remaking of the planet and global society: plastic production rose from 1 million to 300 million tonnes, much of it discarded into waterways to endure for hundreds of years; the number of cars expanded from 40 million in the 1940s to over 1.2 billion today; primary energy use more than quintupled; marine fishing rose from approximately 15 million tonnes to a high of over 70 million tonnes per year; half of global forests were razed; and mankind became the single most important regulator of the chemical cycles governing and sustaining life, namely carbon, sulphur and nitrogen.
> *(Buller 2022:54–5)*

But during the Great Acceleration, 'growth became more than an economic and policy imperative; it became firmly entrenched as an element of ideological hegemony' – 'a fundamental component of the mental infrastructure of the majority of the population of the Western world' (Klein 2024:111).

Within the ideological matrix of post-war hegemony, the obsession with growth interacted with and reinforced the main geopolitical trope, namely Cold War liberalism. In the American hegemon, Cold War liberalism gained supremacy in political culture, in part through the campaign of repression and collective paranoia led by Senator Joe McCarthy from 1950 to 1954. Before the Cold War 'liberalism largely served as an apologia for *laissez-faire* economic policy, and it was entangled in imperialist expansion and racist hierarchy around the world' (Moyn 2023:2–3). However liberalism's hegemonic vision also venerated the free, creative individual and a progressive view of history. Cold War liberalism broke fundamentally from that vision. With the Red Scare pervading political discussion, Marxism, which with its alternative 'vision of a free and equal future, had once been a prompt for liberals to challenge their historical complacency' (Moyn 2023:5) was now public enemy number 1. In the heart of US empire, 'liberalism's relationship to emancipation and reason...disintegrated' and the belief in an emancipated life was now seen as 'proto-totalitarian' (Moyn 2023:4), and worthy of political repression. Within the Cold War Liberal project, as Moyn recounts,

> It was most important to preserve existing liberty in a vale of tears; it was brittle and fragile and always on the verge of assault or collapse. Where earlier liberals had come to accept democratization, if cautiously and often grudgingly, Cold War liberals abhorred mass politics – including mass democracy.
> *(2023:4)*

The anti-communist hysteria was most intense at imperialism's American epicentre, but it radiated throughout the developing Western alliance. Cold War liberalism left a long tail, as precursor to the dual forms of pro-capitalist ideology that claim most of the political spectrum in the United States today: neoliberalism and neoconservatism (Moyn 2023:5).

Fordism and the Allure of Consumer Capitalism

The Great Acceleration of carbon emissions after World War 2 was driven by a set of political-economic developments that birthed a new regime of accumulation, known as Fordism, which in enlarging the historical bloc supporting Fossil Capitalism stabilized the emerging order. Antonio Gramsci first noted these developments, in embryo, in his notes on 'Americanism and Fordism' (1971:285). He recognized in Henry Ford's adoption of mass-production techniques and a high-wages policy, and related efforts by Ford and other American capitalists to control the 'morality' of their workers, a new sort of hegemony, 'born in the factory.' Ford's strategy used the massive productivity increases from reorganizing production to leap ahead of the competition while creating a more stable, loyal and reliable workforce. In Gramsci's view, Fordism blended force and persuasion, to make 'the whole life of the nation revolve around production' (1971: 285). Gramsci's notes 'anticipated the rise of a social structure of accumulation which would provide a basis for the organization of consent' in the post-war era (Carroll 1990:394).

Subsequent analyses have built upon Gramsci's germinal insights. Becker and Weissenbach (2020) clarify that the era of the New Deal, in the late 1930s, was Fordism's period of incubation, as American political leadership came to embrace the Keynesian insight that major economic crises could be avoided by state intervention to manage the economy by keeping aggregate demand for goods and services buoyant. Capitalism's structural tendency towards over-accumulation relative to the collective buying power of consumers could thereby be forestalled. The initial productivity gains from Henry Ford's introduction of the assembly line in automobile manufacturing (subsequently adopted by other industrialists) provided a basis for the relatively high wages Ford paid his workers – creating a loyal workforce with sufficient funds to afford a Model T. After a lengthy period of incubation, New Deal policies in the United States 'were followed by Keynesian state interventionism in nearly all developed Western capitalist countries' (Becker and Weissenbach 2020:114). Overall, the organization of capitalism converged as 'a broad agreement prevailed among the Western democracies, in theory and often in practice, that economic planning, deficit financing, and full employment were inherently desirable and mutually sustaining…' (Pizzolato 2023:36). This system promoted an economic buoyancy that lasted into the 1970s.

Fordism's distinctive combination of mass production for mass consumption transformed capitalism throughout the global North. Its 'productive aspect' transfigured workplaces and workers, converting skilled, craft labour into deskilled, mechanized production and expropriating the know-how of workers – thereby creating unprecedented growth in productivity and in the consumer-goods sector. Fordism's 'consumption aspect' involved 'continual adjustment of *mass consumption* to rises in productivity', achieved through

Keynesian demand management, state-sanctioned collective bargaining and social insurance. In the process, working-class lifestyle was incorporated into capitalist accumulation itself (Lipietz 1987:36). The Fordist historical bloc 'was constructed around a mass producer who was also a mass consumer' (Carroll 1990:397).

As more and more elements of subsistence became commodified, the integration of proletarians into a fully capitalist way of life advanced further. An expansive historical bloc in the global North came to include high-wage segments of the proletariat, whose interests were framed within an apparently stable framework of Fordist mass production for mass consumption, Keynesian management of accumulation to maintain effective demand and welfare-state reforms (Aglietta 1979). Best exemplified by the United States, the long wave of robust accumulation was a time of class compromise and consumerism, cementing a historical bloc around democratic capitalism in the centre. There, the Fordist growth paradigm produced a consensus 'around a shared set of concepts and goals that guided self-styled rational policies geared toward overcoming social conflicts by raising national incomes and strengthening welfare states' (Schmelzer 2016:265). Yet on the periphery, the paradigm entailed neocolonialism via the Open Door of US hegemony and intensified extractivism (Mitchell 2011). The cumulative result was exponentially accelerating carbon emissions – the Great Acceleration. Powered by buried sunshine, the Fordist regime 'was anything but environmentally friendly: high economic growth led to massive increases in CO_2 emissions' (Hope 2018:570; Buch-Hansen and Nielsen 2023:355).

The class compromise at the heart of Fordism introduced a new hegemonic identity – the consumer – which retains its ideological power today. John Watkins (2022) lays out the basic logic of consumer capitalism:

> Capitalism must grow, growth driven by the drive to accumulate capital, propelled by continuous-mass production using fossil fuels. The products churning out from factories, first from Britain and the United States and later from the world, resulted in an abundance of goods for those who could afford them. The output pouring forth led to a cultural change, elevating consumption as the purpose of human existence.
>
> Continuous-mass production and the drive to accumulate profits has molded our culture, habits, and values. It justifies the enjoyment of some at the expense of others. It feeds profits and satisfies our desire for more, all resting on burning fossil fuels, the extraction of which also rests on continuous-mass production. We depend on growth to create jobs, provide economic security, and generate revenues for governments. The affluent accumulate profits; the rest of us accumulate the stuff that the affluent sells.
>
> *(2022:323–4)*

The common-sense of consumerism has, particularly in our era of post-Fordism (to be taken up in the next chapter), displaced the figure of the citizen as a universal subject; indeed, for many people 'democracy' is more a matter of voting-with-your-dollars than participating in collective decision-making. Agency and desire are harnessed into the reproduction of consumer capitalism, creating a particularly strong basis for hegemony. 'Consumerism, as the most recent manifestation of hegemonic domination, offers a false promise of material abundance and enjoyment while producing docility without the use of physical force' (Gotham and Krier 2008:173). With the promise of abundance for all, consumer culture obscures capitalism's relations of exploitation and alienation (Gotham and Krier 2008:182). Not surprisingly, most politicians, intellectuals and citizens 'remain committed to consumerism as an ideology and way of life' that offers an excess of private consumption goods and a dearth of public goods (Schor 1996:505).

As a pivotal element of hegemony, consumerism has powerful social-psychological ramifications. It urges us 'to acquire ever-greater wealth to acquire ever greater consumption', as 'happiness becomes related to the constant growth and expansion of wealth' (Miller 2015:95). Dutifully taking up a position on the treadmill of production seems to be the route to happiness, yet since the commodities we purchase are positional goods, 'the more one has, the more one's neighbours have, the more one wants.' Happily-ever-after thus becomes 'a perpetually receding dream' (ibid.). Ironically, although the acquisitive values of consumerism serve to reproduce consumer capitalism, 'they have negative consequences for individual well-being and sustainability', as privatized lifestyles trap us in the 'iron cage of consumerism' (Isham et al. 2022:16).

Relatedly, the focus on individual consumption weakens working-class solidarities, as worker/consumers compete among themselves for the jobs that enable the golden purchase. As Michael Lebowitz (2020:111) emphasizes, 'for the atomized worker, all other workers are competitors; all other workers are enemies insofar as they are competing for the same jobs. All other workers potentially stand between them and the satisfaction of their needs.' For the atomized worker 'there is always one more opportunity, always some who are worse off to compare with, and always a long way remaining to the top' (Korkotsides 2007:12). Given all this, George Baca (2021:536) characterizes the Fordist era as one of working-class co-optation, disorganization and ideological integration.

> Rather than a compromise, this new matrix of power supplemented an expanding domestic security apparatus with social reforms that resulted in a dialectic of repression and reform. The government subjected the working classes to new forms of control, making labour unions weaker and more vulnerable to corporate power, including the export of capital and

> dismantling labour protections that would follow the late 1970s. These policies steadily weakened organized labour, and more problematically, transformed its desires through mass consumption and celebration of capitalist affluence.

If under the sign of Fordism the marriage of Fossil Capitalism and liberal democracy was consummated across the global North, the implications for climate politics were dire. Éric Pineault's (2021) analysis of the Fordist 'compromise' from a socio-ecological perspective is illuminating on this key issue. Ecologically, the enormous expansion of fossil-fuelled industrial output drove the Great Acceleration in greenhouse gas emissions. The accumulation process 'put to work' biogeochemical processes and ecological relations in production processes that transformed matter/energy into an expanding plethora of commodities, displacing and degrading ecosystem vitality. Yet in the consciousness of both capitalists and workers, biophysical forces 'are turned into ghosts that only appear in short moments of dread' (Pineault 2021:281).

Drawing on Alf Hornborg's (2013) concept of *machine fetishism*, 'a cultural tendency in capitalist economies and societies to attribute to technology productive capacity while obscuring the biophysical origin of technological systems, as if capital fixed as technology was inherently productive' (2021:270), Pineault considers a critical socio-political implication of the Fordist accord between capital and labour. From a socio-ecological perspective, the accord created a stable structure in the advanced capitalist core which was achieved 'by a precise form of displacement.' Pineault explains: 'an inner contradiction between capital and labour is stabilized and mitigated by creating an exterior contradiction between capital and nature: climate change through the combustion of fossil fuels to energize the productive and consumptive process and thus drive aggregate growth' (2021:280). In this account, as workers in Fordism's heyday were able to capture through collective bargaining a share of productivity gains, each gain contributing to rising profits and wages was achieved, on the treadmill of production, by depleting and degrading ecosystems.

Here is where machine fetishism enters the scene. Pineault observes that

> both parties to the Fordist compromise are subject to machine fetishism because of their constituent class position in the labour process. They will see the expansionary principle of capitalist accumulation during the mid-20th century as a natural expression of its social relations, and they will see the materialities involved as contingent: social causality veiling material causality. Climate change and other ecological contradictions are just unfortunate consequences. Furthermore, because their class interests are embedded in the narrow labour process of machine

> fetishism, the constituent classes of the deal will also see in capitalist expansion and productivity growth the solution to the 'after the fact' environmental problems. They will opt for green growth.
>
> *(2021:280)*

Pineault's analysis gives fresh meaning to Gramsci's assessment of Fordism: 'hegemony here is born in the factory' (Gramsci 1971:285). Clearly, the implications of a machine fetishism shared by both parties to capitalist production are still with us today, as a powerful element organizing consent to Fossil Capitalism. For workers now fully ensconced in capitalism's relations of production and reproduction, the struggle appears to be one of wresting further productivity gains in the form of higher wages, without comprehending the internal relation between their own alienated labour and the perilous degradation of nonhuman nature. It is no wonder that today, few workers, consumers and citizens in the global North can see past 'green growth' as a solution to the ravaging climate crisis.

Lifeblood: Fossil Capitalism Becomes Common Sense in the Global North

As we have seen, in the era of Fordism persuasive means of hegemonic incorporation developed within Fossil Capitalism, which operated both economically and ideologically – a faith in 'growth' (aka capital accumulation), the Cold War liberal othering of alternatives to capitalism as unrealistic and authoritarian, the predominance of consumerism and the integrative force of Fordism. Today, these retain much of their persuasive power decades after the crisis of Fordism that led to the current regime of financialized neoliberal capitalism (now also in crisis).

In an astute case study of the United States as epicentre of this way of life, Matthew Huber (2013) has explored how, within the everyday lives of Americans, the hegemony of fossil capital was consolidated in the post-war era. Huber asks, how did 'oil' become central within the construction of an 'American way of life' (Huber 2013:xv)? His research is particularly revealing in showing how dominant social arrangements introduced under Fordism set the table for the rise and popular acceptance of post-Fordist neoliberal capitalism. Just as the period of the New Deal and World War 2 incubated the Fordist regime of accumulation, Huber argues that

> the postwar period must be viewed as neoliberalism's *incubation period* wherein popular resentment of government, taxes, and Keynesianism festered and built itself until the political moment of opportunity in the 1970s emerged— a moment structured in no small part by concerns over 'oil shocks.' In fact, the historical roots of neoliberalism stretch back to the

> contradictions of the New Deal project itself (if not much earlier). Although New Deal liberalism was based around collective narratives of public solidarity and the beneficial role of state intervention, it created the conditions for a privatized geography of suburbanization.
>
> *(2013:xvi)*

Huber begins his analysis by pointing to the widely acknowledged problem of 'oil addiction' which, most agree, 'is rooted in the particular form of autocentric, suburban development prevalent in the United States' (2013:x). Rejecting the implication that addiction implies an pernicious entity within America's bloodstream, he asks, 'What if the most problematic relation to oil is the way it powers forms of social life that allow individuals to imagine themselves as severed from society and public life?' (2013:xi). Huber's historical materialism views petroleum as a material aspect of the broader power of capital over living labour, as 'a central energy resource shaping the forces of social reproduction' in 'the real subsumption of life under capital', through which 'life *appears as capital*'(2013:xiv). Like Pineault, Huber sees in this subsumption a process of fetishization: 'it is historically specific to capitalism for social change to *appear* as the product of fetishized things – machines, technology, oil – rather than social labor' (2013:12). It is in the Fordist era that life became really subsumed under capital, as the reproduction of workers' lives was incorporated, as consumer capitalism, within capital's economic circuitry. In this process, 'powered not through muscles through but fossil-fuel combustion' (Huber 2013:18), oil became *lifeblood* for a distinct way of life now known as Fossil Capitalism.

The unfolding throughout the 20th century of mass production for mass consumption brought the industrial revolution into 'the reproductive forces of everyday life' (2013:16), for substantial segments of the American working class. For them, the geographies of everyday life were transformed in three ways. First, the new lifestyle centred upon the single-family home, detached from other residences and sited on land owned by the homeowner – a scenario quite different from that of the 19th-century propertyless proletariat. Second, this new geography brought dispersed property into 'low-density suburban settlements characterized by an abundance of privatized spaces for privatized families.' Third, for suburbanites to traverse the vast new spaces 'a new privatized and commodified form of transport' was required, namely, automobility (2013:17). Huber's study shows how 'this particular spatiality of single-family homes, cars, yards, and highways came to be constructed as a specifically "American way of life"' (ibid.), as it became generalized after World War 2 primarily among the more affluent, mainly white male workers and their families. Suburban homeowners surfaced as 'a specific "historical bloc"', essential not only to the organization of consent in the Fordist era but, to the later victories of neoliberalism, securing 'popular consent to neoliberal logics and common sense' (2013:20).

The New Deal reforms that endeavoured to create an American way of life based in high wages and purchasing power for a segment of the proletariat, and the resulting class compromise (higher wages for industrial peace), were only achieved through a profound 'ecological compromise.' 'The very notion of "mass" consumption presupposes a mass of materials, energies, and wastes channeled into the geographies of social reproduction and the cultural politics of "the American way of life"' (Huber 2013:42). Key to really subsuming life under capital was 'the construction of a realm of freedom subsumed in commodity relations' (Huber 2013:87). The suburban home became that realm, a private space, connected to the wider world through fossil-fuelled automobiles and powered by a host of household appliances and by food made cheap by a fossilized production system (Huber 2013:87). Corporations manufacturing these products (with petroleum producers in the lead) played an important ideological role through their advertising campaigns.[3] In the vision of an oil-fuelled way of life, spaces of leisure and privatized transportation are identified with freedom, in opposition to the realm of work which implicitly becomes an 'interruption of one's real life of family, home, and leisure' (Huber 2013:86).

As this oil-fuelled way of life became solidified, the fit, within Fordism and consumer capitalism, between accumulation regime and hegemonic project tightened into a predominant common sense.

> In effect, some parts of the working class were energized, afforded enormous power over machines, space, and everyday life in navigating the practices of reproduction. Critically, this was a specific form of energy and power – privatized power, individuated command over space in the automobile and a veritable mechanized factory of reproduction in the household.
>
> *(Huber 2013:159)*

Within this way of life, 'the individual experiences automobility as empowering and liberating, and the single detached house as a domain of personal sovereignty' (Carroll 2021:6). This is the life that many middle-class folks in the global North now lead (including me!). It is not a way of life that could be generalized to all of humanity, and it is not ecologically sustainable, as we are now seeing.

Huber's case study cannot be automatically generalized across the advanced capitalist democracies. The real subsumption of life within Fossil Capitalism occurred in varying degrees throughout the global North during the Fordist era. Certainly, the American case is the purist in this regard, as the United States' hegemonic position within global capitalism, the remarkable strength and duration of its post-war boom economy and the eradication of the socialist left during the Cold War 1950s rendered the working class 'prisoners of the American Dream', in Mike Davis's (1986) famous phrase. Suburbanization (along with automobility) was most extensive in the settler colonies of the United States, Australia and Canada – where an abundance of land, stolen from Indigenous peoples, could be easily accessed. Also, in countries where a strong left current persisted, grounded in a robust labour movement (typically in the form of social democracy), the tendency for privatized geography of everyday life to feed atomized individualism was tempered. Still, as cars, TVs and corporate advertising, appliances, petroleum products and the private home came to saturate everyday life the suburbanite was recruited to support the endless growth of carbon-driven consumer capitalism.

A longer-range political ramification was the constriction of politics within narrow limits focussed not only on 'growth' but 'on the family, private property, and anticollectivist sentiments' (Huber 2013:79) – the stock-and-trade of the neoliberalism that would come to supplant the Fordist-Keynesian project in the 1980s, just as the climate crisis was becoming more visible. I will trace these development in the next chapter. At this point, we can note that oil's ascension to the lifeblood of consumer capitalism tightened the grip of fossil capital on accumulation and on hegemonic understandings of prosperity. In the post-war Zeitgeist, '"economic growth" was in fact just another name for the petrolization of the world' (Fressoz and Bonneuil 2017:60).

The Imperial Mode of Living

Matthew Huber's case study is centred on the country that has served as ground zero for the carbon bomb that is contemporary Fossil Capitalism. Although he acknowledges the imperialist relations that were always integral to Fordist consumerism as a way of life within advanced capitalism,[4] Huber's focus is on the American experience. Ulrich Brand and Markus Wissen's (2021) study of *The Imperial Mode of Living* complements Huber's analysis in bringing the third aspect of our theoretical trifecta – accumulation, hegemony, imperialism – into focus, thereby illuminating how the 'American way of life' became an imperial mode of living, global in scope.

Brand and Wissen trace the imperial mode of living from its colonial beginnings in early capitalism, when only the ruling class could enjoy affluence in its everyday life, through the Fordist era in which the more affluent sections of the Northern working class came to imitate upper-class lifestyles within the social logic of consumer capitalism, to its 'generalized application today' (2021:10). Such imitation rested on the productivity gains from adoption of mass-production methods, which funded higher wages. The Northern working class could have instead struggled for a shorter work week, and more available free time, but the class compromise was founded on increases in working-class consumption that sustained expanding markets and profit opportunities for capital (2021:89–90). The compromise brought relative stability and industrial peace in the advanced capitalist core countries, as labour took up an ambivalent partnership with capital. Centred in the advanced capitalist countries, this mode of living is *imperial* for a compelling reason. It 'depends upon the worldwide exploitation of nature – and wage and non-wage labour – while simultaneously externalizing the social and ecological consequences arising from it' (2021:4).

Brand and Wissen's aim is to understand how, within the global North, the 'normality' of consumer capitalism 'is produced precisely by masking the destruction in which it is rooted' (2021:5). Although the imperial mode of living causes suffering and destruction elsewhere, for many in the global North (and increasingly for middle classes in the global South) this way of life 'means the opportunity to have a subjectively fulfilled life', enhanced by 'home appliances, industrialized food, cars, smartphones' (2021:53). As we saw in Chapter 1, within eco-imperialism the global South serves both as 'tap' and 'sink' for the treadmill of production – thereby requiring its continued subordination (both people and nonhuman nature) to the requirements of capital accumulation in the North. In short, the advance of Fordism meant that externalization would become an integral condition for mass consumption/production in the global North. And the more these emissions and resource-intensive practices spread across the North the greater the need for an elsewhere, to which socio-ecological costs could be shifted and from which resources could be sourced. At the level of popular consciousness, NIMBYism (Not in my Backyard) became a normalized attitude among consumer capitalism's atomized subjects, expressing the common sense of externalization (2021:96). Thus,

> the imperial mode of living is based on exclusivity; it can sustain itself only as long as an 'outside' on which to impose its costs is available. But this 'outside' is shrinking as more and more societies access it and fewer people are willing or able to bear the costs of externalization processes. The imperial mode of living is thus becoming a victim of its own appeal and universalization.
>
> *(Brand and Wissen 2021:7)*

Indeed, that the crises and contradictions we now face – ultimately, the prospect of ecocide – are intensifying today 'can be attributed to the fact that this mode of living is in the process of succeeding even at the cost of self-destruction' (2021:6).

Huber's study of oil as lifeblood and Brand and Wissen's analysis of the imperial mode of living illuminate the distinct configuration within which our trifecta of capital accumulation, imperialism and hegemony combined, during the Great Acceleration. Within Pax Americana their combination yielded a form of capitalism that offered unprecedented affluence to a segment of the Northern proletariat, yet relied on resource imperialism and an ever-escalating carbon footprint. Within the capitalist democracies, high-wage fractions of the proletariat were economically and ideologically incorporated into an expanding historical bloc, through the everyday practices of consumerism, automobility and suburbanization. In the international field, the post-war reconfiguration of imperialism established conditions for not only US hegemony over its allies in the Western bloc and neo-colonies in the global South, but for the generalization of Fordism as a regime of accumulation throughout the 'West' in a transnational historical bloc unified through the imperial mode of living.

All the while, global carbon emissions expanded exponentially, increasing by 300 percent from 1945 to 1975 (Ritchie and Roser 2024), but the increasingly carboniferous atmosphere did not yet engender the symptoms of climate breakdown we now experience.

The Crisis of Post-war Fordism

As I emphasized at the start of this chapter, capitalism is a crisis-prone, crisis-ridden and crisis-dependent form of society. The rhythm of accumulation follows a boom/bust cycle, both in the medium and long term. By the 1970s, the long wave of Fordist accumulation that underwrote the full flowering of the post-war accord had begun to break down. A number of developments contributed to the crisis.

Considering the situation from the side of industrial capital, the innovations in production that Fordism instigated had run their course by the late 1960s, attenuating the productivity gains that funded class compromise. To complicate matters further, state recognition of trade unions as legitimate collective representatives of their members – a core aspect of the compromise – had enabled unions to gain strength, and the Keynesian emphasis on keeping aggregate demand buoyant through policies of 'full employment' enhanced labour's bargaining position. All these contributed to a profit squeeze, leading capitalists to look towards more promising investment outlets. The transnational corporations that had grown enormously during the post-war boom increasingly began to offshore manufacturing to the South to take advantage of low wages and light regulation. By the 1970s this new international division of labour was contributing to industrial stagnation in the core economies, symptomatically registered in worries about deindustrialization and in an emerging 'rust belt' in the American mid-west.[5] In the United States and other core economies, 'stagflation' loomed as industrial capitalists, reluctant to invest in production, protected their own profits by passing on the cost of wage increases to consumers, stoking inflation.

Considering the situation from the side of financial capital, the post-war, American-led order had opened borders not only for transnational corporations but also for the international flow of financial capital. Stimulating this was the invention of new financial instruments such as hedge funds and derivatives and the consolidation of a global financial market, connected through 24/7 exchange trading (McNally 2011). This incipient financialization unleashed vast opportunities for capitalists to realize profits without tying up their capital in physical production. The result was 'the hypertrophy of ("stateless") money capital in the international circuit of capital' (van der Pijl 2012:259), which sharpened rivalries among the imperialist powers while augmenting the relative power of finance within accumulation – prefiguring the 'casino capitalism' that would explode onto the scene in the 1980s (Strange 1986).

Kees van der Pijl (2012) reminds us that the structural division of capital into industrial and financial forms creates a *fractional division* within the capitalist class, which carries portentous implications for the shape and form of bourgeois hegemony. Industrial capital (especially large-scale production requiring enormous investment in factories and infrastructure) has an interest in the technical and social arrangements in commodity production, which entail site-specific planning, labour relations, etc., to capture a competitive share of total surplus value in the form of industrial profit. Financial capital's immediate interest is in the free flow of money-capital and in 'sound money' that retains its value, enabling the capture surplus value from debtors and (increasingly) from the buying and selling of financial instruments. Despite the strong tendency for industrial and financial capital to become integrated, as 'finance capital' in the era of monopoly capitalism and imperialism (Hilferding 2006), 'certain conflicts within the capitalist class remain traceable to the different fractions persisting in the context of an apparent fusion' (van der Pijl 2012:7).

In the Fordist strategy, financial capital was subordinated to productive capital in each national economy, but also embedded within a dollar-based international financial system that promoted a regulated form of capitalist internationalization, through reliable currency convertibility. By the 1970s, declining competitiveness had diminished the American commitment to the burdens of leadership this arrangement implied. In 1971, the United States abrogated the Bretton Woods Agreement of 1944, which had directly aligned the dollar with a stable price for gold. With the associated rise of markets in offshore dollars (Eurodollars), financial capital was effectively liberated from regulation by central banks. Its unimpeded international circulation promoted the transfer of North Atlantic industry to the South, with telling consequences (Carroll 1990:399). In breaking the territorial coincidence between mass production and mass consumption, the new international division of labour subverted both the class compromise and the very structure of Atlantic integration at the heart of the post-war historical bloc (Van der Pijl 2012: xviii).

The Keynesian Welfare State, which had mediated class compromise by redistributing income and developing a suite of social programmes and services in the field of social reproduction, was increasingly seen in elite circles as a drag on accumulation, as it confiscated funds that could be turned to account as capital. By the late 1970s, as Kees van der Pijl (1986:27) argues, financial capital 'could lead the way towards the reassertion of the freedom of capital from political constraints ... In this sense, it represented an interest wider than its own, for industrial capital too would in the long term prefer a redefinition of its operating conditions in social terms.'

So it was that the Fordist regime gave way to a new accumulation strategy and hegemonic project: neoliberalism. Van der Pijl, and the Amsterdam School of transnational historical materialism he has led (Jessop and Overbeek 2018), point out that the financial/industrial division deeply embedded within capital accumulation offers two distinct perspectives or standpoints from which hegemony can be constructed. In the era of Fordism, the Keynesian Welfare State and rising US hegemony, the standpoint of productive, industrial capital held sway, emphasizing the generalization of mass production and associated industrial and social planning processes as well as the need for a reliable, high-wage workforce.

> With a focus on raising wages to expand effective demand, Keynesian growth policies transformed how the working classes participated in capitalist accumulation, which limited working-class politics to purely economic needs.... Based on this new accumulation model, welfare state policies became productive in making labour central to economics as consumers. In this way, the Keynesian project imposed power upon working classes in ways that made them more docile and increasingly committed to capitalism through consumerism.'
>
> *(George Baca 2021)*

Neoliberalism, in contrast, took up the standpoint of financial capital. The emergent project responded to the crisis of Fordism with a policy framework intended to liberate capital from what were now seen as impediments to accumulation. It championed free labour unencumbered by unions, 'sound money' in place of galloping inflation, unfettered international financial circulation, the elimination of state budget deficits and the privatization of public assets to stoke accumulation. It thus put forth the perspective of financial capital as a *general interest* around which other social interests could be assembled in a hegemonic project (Van der Pijl 1986:3; Overbeek 2024).

As we shall see in the next chapter, the convergence of the rise of neoliberalism and the deepening of the climate crisis is a case of very bad timing Klein 2024 (Klein 2014), as it not only intensified economic crisis tendencies, but made effective solutions to the climate emergency extremely difficult.

Notes

1 'In 1820 the per capita income of the wealthiest countries was 3 times that of the poorest. By 1870 it was 7 times and by 1913 it was 11 times higher. By 1997, the one-fifth of the world's population living in the richest countries was 74 times as rich as the one-fifth in the poorest, up from 60 in 1990 and 30 in 1960' (Dunford and Yeung 2011:23).

2 In 1945 the United States maintained 905 foreign military bases worldwide; by 1989 it maintained 1073 (Vine 2021).

3 Huber displays as an example an Esso (now Exxon) advertisement that claimed, '"An oil discovery that helps you *eat* better!" with images of healthy fruits and vegetables made possible through "a brilliant new chemical ... hailed as one of the most versatile and effective fungicides in existence"' (2013:88).

4 'Through the lens of fetishism, oil is envisioned as a vital and strategic "thing" through which imperial relations are solidified (petro)states are formed, and local livelihoods are violently destabilized. Thus the site of extraction becomes the territorial center through which conflicting social forces congregate and struggle over the oil "prize" unfolds. Oil is not a thing-in-itself. Contrary to fetishistic discourses of "oil states," "oil wars," and "oil addictions," a dialectical approach must seek to understand oil as a socioecological relation. I have proposed a much broader historical-geographical materialist perspective that situates oil in particular, and energy more broadly, within the "production and reproduction of life." Thus political resistance to the geopolitical games of imperial control over oil reserves must cast their critical sights toward not only the U.S. military state but also the geographies of social reproduction that situate oil as a necessary element of "life"' (Huber 2013:25–6).

5 In this period the United States began to cede its position of economic predominance, 'a loss that is due largely to its successful export of the expansive Fordist production and consumption model – particularly to Western Europe: in other words, as a result of the generalization of the imperial mode of living in the global North' (Brand and Wissen 2021:96).

References

Aglietta, Michel. 1979. *A Theory of Capitalist Regulation*. London: NLB.

Baca, George. 2021. "Neoliberalism's Prologue: Keynesianism, Myths of Class Compromises and the Restoration of Class Power." *Anthropological Theory* 21(4):520–40. doi: 10.1177/1463499621989130

Becker, Joachim, and Rudy Weissenbacher. 2020. "Fordism." Pp. 113–18 in *The Routledge Handbook to the Political Economy and Governance of the Americas*, edited by O. Kaltmeier. Abingdon, Oxon; New York, NY: Routledge.

Brand, Ulrich, and Markus Wissen. 2021. *The Imperial Mode of Living*. London: Verso.

Buch-Hansen, Hubert, and Peter Nielsen. 2023. "Critical Realism, the Climate Crisis and (de)Growth." *Journal of Critical Realism* 22(3):347–63. doi: 10.1080/14767430.2023.2217050

Buller, Adrienne. 2022. *The Value of a Whale: On the Illusions of Green Capitalism*. Manchester: Manchester University Press.

Carroll, William K. 1990. "Restructuring Capital, Reorganizing Consent: Gramsci, Political Economy, and Canada*." *Canadian Review of Sociology/Revue Canadienne de Sociologie* 27(3):390–416. doi: 10.1111/j.1755-618X.1990.tb00132.x

Carroll, William K. 2021. "Conclusion: Prospects for Energy Democracy in the Face of Passive Revolution." Pp. 479–504 in *Regime of Obstruction*, edited by W. K. Carroll. Edmonton: AU Press.

Davis, Mike. 1986. *Prisoners of the American Dream*. New York: Verso.

Dunford, Michael and Godfrey Yeung. 2011. "Towards Global Convergence: Emerging Economies, the rise of China and Western Sunset?" *European Urban and Regional Studies* 18 (1):22–46.

Fressoz, Jean-Baptiste, and Christophe Bonneuil. 2017. "Growth Unlimited: The Idea of Infinite Growth from Fossil Capitalism to Green Capitalism." Pp. 52–68 in *History of the Future of Economic Growth: Historical Roots of Current Debates on Sustainable Degrowth*, edited by I. Borowy and M. Schmelzer. London; New York, NY: Routledge, Taylor & Francis Group.

Gotham, Kevin Fox, and Daniel A. Krier. 2008. "From the Culture Industry to the Society of the Spectacle: Critical Theory and the Situationist International." Pp. 155–92 in *Current Perspectives in Social Theory*. Vol. 25, edited by Harry F. Dahms. Bingley: Emerald Publishing Limited.

Gramsci, Antonio. 1971. *Selections from the Prison Notebooks*. edited by Q. Hoare and G. Nowell-Smith. New York: Lawrence & Wishart.

Harvey, David. 2006. *The Limits to Capital*. London: Verso.

Hilferding, Rudolf. 2006. *Finance Capital: A Study of the Latest Phase of Capitalist Development*. London: Routledge & Kegan Paul.

Hobsbawm, Eric. 1989. *The Age of Empire: 1875–1914*. New York: Vintage Books.

Hope, Wayne. 2018. "Epochality, Global Capitalism and Ecology." *TripleC: Communication, Capitalism & Critique* 16(2):562–76. doi: 10.31269/triplec.v16i2.1002

Hornborg, Alf. 2013. *Global Ecology and Unequal Exchange: Fetishism in a Zero-Sum World*. New York: Routledge.

Huber, Matthew T. 2013. *Lifeblood: Oil, Freedom, and the Forces of Capital*. Minneapolis: University of Minnesota Press.

Isham, Amy, Caroline Verfuerth, Alison Armstrong, Patrick Elf, Birgitta Gatersleben, and Tim Jackson. 2022. "The Problematic Role of Materialistic Values in the Pursuit of Sustainable Well-Being." *International Journal of Environmental Research and Public Health* 19(6):3673. doi: 10.3390/ijerph19063673

Jessop, Bob, and Henk Overbeek, eds. 2018. *Transnational Capital and Class Fractions: The Amsterdam School Perspective Reconsidered*. London: Routledge.

Kaya, Muhammet. 2022. "1873–1896 Long Depression." Pp. 57–77 in *Black Swan: Economic Crises, Volume I*, edited by B. Açikgöz. Singapore: Springer Nature Singapore.

Klein, Dieter. 2024. *Within and Beyond Capitalism: A Twofold Transformation*. London: Routledge.

Klein, Naomi. 2014. *This Changes Everything: Capitalism vs. the Climate*. Toronto: Alfred A. Knopf Canada.

Korkotsides, Anastasios S. 2007. *Consumer Capitalism*. London; New York: Routledge.

Lebowitz, Michael A. 2020. *Between Capitalism and Community*. New York: Monthly Review Press.

Lipietz, Alain. 1987. *Mirages and Miracles*. London: Verso.

McNally, David. 2011. *Global Slump*. Oakland, CA: PM Press.

Mandel, Ernest. 1975. *Late Capitalism*. London: NLB.
Miller, Catriona. 2015. "Finding the Golden Egg: Illusions of Happiness in an Age of Consumer Capitalism." Pp. 89–106 in *The Happiness Illusion: How the Media sold us a Fairytale*, edited by L. Hockley and N. Fadina. London; New York, NY: Routledge.
Mitchell, Timothy. 2009. "Carbon Democracy." *Economy and Society* 38(3):399–432. doi: 10.1080/03085140903020598
Mitchell, Timothy. 2011. *Carbon Democracy*. London: Verso.
Moyn, Samuel. 2023. *Liberalism against Itself: Cold War Intellectuals and the Making of Our Times*. New Haven, London: Yale University Press.
O'Connor, James. 1987. *The Meaning of Crisis: A Theoretical Introduction*. London: London: Basil Blackwell.
Overbeek, Henk. 2024. "Transnational Neoliberalism in Organic Crisis." Pp. 428–47 in *The Elgar Companion to Antonio Gramsci*, edited by W. K. Carroll. Cheltenham, UKS; Northampton, MA: Edward Elgar Publishing.
Pijl, Kees van der. 1986. *Neoliberalism vs. Planned Interdependence. Concepts of Control in the Struggle for Hegemony*. Groningen, Netherlands: Polemologisch Instituut Groningen.
Pijl, Kees van der. 2012. *The Making of an Atlantic Ruling Class*. Second Edition. London: Verso.
Pineault, Eric. 2021. "The Ghosts of Progress: Contradictory Materialities of the Capitalist Golden Age." *Anthropological Theory* 21(3):260–86. doi: 10.1177/1463499620980292
Pizzolato, Nicola. 2023. *Challenging Global Capitalism*. London: Palgrave Macmillan.
Ritchie, Hannah, and Max Roser. 2024. *CO_2 Emissions*. Our World in Data. Published online at OurWorldInData.org. Retrieved August 27, 2024 (https://ourworldindata.org/co2-emissions).
Schmelzer, Matthias. 2016. *The Hegemony of Growth: The OECD and the Making of the Economic Growth Paradigm*. Cambridge, UK: Cambridge University Press.
Schor, Juliet. 1996. "What's Wrong with Consumer Capitalism? *The Joyless Economy* after Twenty Years." *Critical Review* 10(4):495–508. doi: 10.1080/08913819608443435
Strange, Susan. 1986. *Casino Capitalism*. Oxford: Basil Blackwell.
Vine, David. 2021. "Lists of U.S. Military Bases Abroad, 1776–2021."Washington, D.C.: American University. Online resource. https://doi.org/10.17606/7em4-hb13
Watkins, John P. 2022. "The Origins and Evolution of Consumer Capitalism: The Paradoxes Posed by Continuous Mass Production." *Journal of Economic Issues* 56(2):314–25. doi: 10.1080/00213624.2022.2050140

3

CLIMATE CRISIS AND THE QUICKENING OF FOSSIL CAPITALISM'S DEATH DRIVE

In this book's Introduction, I laid out the symptoms of climate breakdown, which follow closely from the exponential rise in carbon emissions. It is a bitter irony that emissions made their greatest jump precisely in the era when knowledge of the link between burning carbon and heating the planet became unassailable. Of the world's 1.77 trillion tonnes of energy-related carbon dioxide emissions that have been released over the course of human history, half have occurred in the past 30 years (Nikiforuk 2024). In 1988 James Hanson, then Director of NASA's Goddard Institute for Space Studies, gave his famous testimony before American Congress, explicitly drawing the causal link between carbon emissions and global warming. Less than a decade later, the Kyoto Protocol was adopted at the third Conference of the Parties (COP) in 1997, extending the 1992 United Nations Framework Convention on Climate Change. The Protocol, which entered into force in 2005 (and which the United States refused to ratify) mandated emission reduction targets amounting to 'an average 5 per cent emission reduction compared to 1990 levels over the five year period 2008–2012' (United Nations n.d.). Yet, as I noted in this book's Introduction, the planet continues on a course towards 'hothouse earth.' In this chapter, I retrace the steps, from the 1980s, that have taken us to our current quandary.

Neoliberalism and the Climate Emergency: Bad Timing

As climate chaos gained visibility in the 1980s and 1990s, the policy environment within which Fossil Capitalism moves shifted dramatically. The coincidence, over the past four decades, of neoliberal ascendency and increasingly dire climate breakdown presents a case of very bad timing (Klein 2014).

DOI: 10.4324/9781003412922-5

Neoliberalism emerged in the late 1970s as a political-economic paradigm intended to resolve the crisis of Fordism. Its implementation in the 1980s in much of the global North and its generalization throughout the world in the 1990s unfolded as the climate crisis began to bite. For neoliberals, 'free' markets 'are the most moral and the most efficient means for producing and distributing goods and services' (Cahill 2012:111); hence a society in which 'the market' guides economic production and distribution is ideal. As the climate crisis worsened, political leadership set about 'freeing' markets in the name of efficiency and liberty, on the dubious premise that the rationality of the 'invisible hand' of inter-capitalist competition would produce climate solutions.

Most professional economists (to say nothing of mainstream politicians) accept this premise, which has created 'a wide and enduring gulf between economists and scientific reality when it comes to the climate and nature crises' (Buller 2022:41). For instance, a recent economic study (Dietz et al. 2021) concluded that in a world six degrees Celsius warmer than today's, global per capita consumption would shrink by a measly 1.4%. This projection, couched in the language of GDP, not human welfare, contrasts sharply with the consensus among climate scientists on ecological tipping points and the catastrophic implications of a six degree rise in average temperature. 'The combined impact of widespread permafrost thaw, the cessation of vital ocean currents, a net-emitting Amazon rainforest and the loss of the Greenland ice sheet – to name a few – together constitute an utterly unimaginable future in which planetary conditions would resemble those not seen for tens of millions of years' (Buller 2022:41). The absurdity of projecting a 1.4% drop in 'global consumption' in the context of such ecological disaster is plainly obvious, to anyone but a liberal economist! Yet as we shall see in Chapter 4, solutions to the climate crisis emanating from the halls of political and economic power stay entirely within market-centred and technocratic frameworks inured to the gravity and urgency of the climate emergency.

Although the first neoliberal regime was established as a police state in Chile, through a US-backed military coup that brought General Augusto Pinochet to power in 1973, neoliberalism's ascendance as a hegemonic project is typically dated at the elections of the Margaret Thatcher (1979) and Ronald Reagan (1980) governments, respectively, in the UK and the United States. Neoliberalism 'proposes that human well-being can best be advanced by liberating individual entrepreneurial freedoms and skills within an institutional framework characterized by strong private property rights, free markets, and free trade. The role of the state is to create and preserve an institutional framework appropriate to such practices' (Harvey 2005a:2). Under neoliberalism, the state guarantees the premises of capital accumulation – sound money, the institutions that defend private property rights (police, law) and the proper functioning of markets. But 'beyond these tasks the state should not venture' (ibid.). David Harvey goes on to argue that 'neoliberalization was from the very beginning a project to achieve the restoration of class power' (Harvey 2005a:16), in the wake of the failure of Fordism to maintain both class compromise and robust accumulation. In this, the project was a resounding success; for instance, in the United States, the ratio of median wages for workers to CEO salaries jumped from 30 to 1 in 1970 to 500 to 1 30 years later (ibid.).

To understand neoliberalism as a class project we can recall, with Gary Teeple, the actual meaning of private property rights within a world dominated by giant corporations. Private property 'implies the exercise of power over things produced, and even over the means of producing things' (Teeple 2008:82). But there is a vast difference in the exercise of power between, say, owning a car and house (the standard personal effects of Fordism) and owning controlling interest in a corporation. In the latter case,

> because capitalist private property entails exclusive corporate ownership and control over the means of production, which in turn denotes neither ownership nor control by those who must work for another to live, such property rights are the very basis of economic inequality and its continual increase.
>
> *(Teeple 2008:83)*

In the analysis of surplus value in Chapter 1, we saw how this process works, and how, historically, as capital accumulates it concentrates and centralizes into fewer and fewer, larger and larger units – the giant corporations and financial institutions that now rule the world. As a hegemonic project, neoliberalism presented the 'free market' as the desirable alternative to Fordist state programming and regulation, yet 'given the degree of cartelization, monopolization, and oligopolization, the reality of the market is to a very large degree an illusion and has been for some time' (Teeple 2008:86). Bluntly, in a setting of concentrated corporate power, the 'freeing' of markets has the effect of further empowering the capitalists who control economic resources. Moreover, there can be no 'return' to the 'free market': 'the direction of movement is towards the minimization of competition and maximization of control over all markets, and these trends are the natural outcome of capitalist development' (Teeple 2008:87).

Recall from Chapter 2 that in the shift from Fordist-Keynesian to neoliberal policy the standpoint from which capitalist hegemony is constructed also shifted, from a conception of capitalism revolving around productive capital to a conception emphasizing the perspective of financial capital. The shift has also been described as one from 'demand-side' to 'supply side' policy. Keynesian policy had endeavoured to 'manage' capitalism by addressing the chronic problem of underconsumption: as capitalists compete with each other in the quest for profit they press wages down, but an immiserated working class cannot purchase the commodities the system produces. Lack of 'effective demand' – particularly working-class purchasing power – creates a glut of unsaleable commodities, discouraging further investment. The theory was that relatively high wages and expansive social programmes would keep aggregate demand buoyant, assuring industrialists of demand for their products and thereby inducing accumulation. But as the Fordist long wave crashed in falling profit rates (Carchedi and Roberts 2018), accompanied by state deficits and hyperinflation, the Keynesian policy paradigm lost favour. Neoliberal supply-side logic gained persuasiveness in mainstream political circles.

Viewed from the supply side, the key to sustained and robust accumulation is the supply of capital, i.e., capitalist investment. To 'free' markets, to stoke a new round of accumulation, neoliberal policy implemented several 'supply-side' measures to improve profit prospects for investors. Neoliberal regimes prioritized 'sound money' to assure financial capitalists that their assets would retain value rather than shrink through inflation. They attacked organized labour – a collective power generating the high wages that were squeezing profits. They attacked state deficits (financed by banks) that were diverting capital from private-sector investment, cut social programmes that were deemed to have a similar crowding-out effect and also cut taxes (particularly on the wealthy) that had funded those programmes but were now seen as a drag on investment. They deregulated capital by removing many rules and regulations meant to protect communities and curb arbitrary corporate power (including environmental regulations). Internationally, neoliberal deregulation emphasized 'free trade' agreements and World Trade Association measures to expand the powers of foreign investors, as capitalism became more globalized and financialized. These policies *insulated* capital from democratic decision-making. As the KWS was hollowed out and as regulations on business were loosened, corporate capital gained more freedom to invest according to its priorities. The neoliberal state itself also became more insulated from its citizenry, as central banks gained 'autonomy' from elected governments while technocrats in the WTO, World Bank and other quasi-state organizations facilitated the opening of new fields for accumulation (Carroll 2006:13). Importantly, the weakening of the KWS did not diminish state power; rather, it shifted state strategies from the hegemonic management of a class compromise to the defense of capital's expanded rights against now-disreputable social forces – organized labour, welfare recipients, the left. As the increasing militarization of policing and criminalization of dissent shows, since its inception neoliberalism has combined the 'free economy' with a 'strong state' enforcing property rights against resistance from below (Gamble 1994).

Finally, neoliberal policy *privatized* public and other commonly held assets, transferring ownership of utilities and other public goods to capitalists.

Harvey (2005a) identifies these practices as 'accumulation by dispossession', and locates them at the centre of a 'new imperialism.' While neoliberal insulation of capital from democratic constraint promotes accumulation within liberalized markets, accumulation by dispossession converts commonly held assets (or rights to assets) into new venues into which surplus capital can flow. In releasing these assets to capital, typically at low cost, accumulation by dispossession opened new markets, driving a spectacular increase in financial speculation. This accumulation strategy, an aspect of cannibal capitalism (Fraser 2022) has taken many forms – 'biopiracy and the wholesale commodification of nature, commercialization of culture and intellectual creativity, corporatization and privatization of public institutions and utilities – in short, the enclosure of the commons' (Carroll 2006:14). Harvey's point is that the neoliberal project brought with it a new imperialism, as the 'privatization of everything' (Harvey, 2005b:149) created an impetus 'to find more and more arenas, either at home or abroad, where privatization might be achieved' (Harvey 2005b:158).

If neoliberalism shifted the hegemonic standpoint from nationally based industrial capital to internationally mobile financial capital, while concurrently shifting state economic priorities from aggregate demand management to incentivizing the supply of capital, a third shift is worth recalling at this point. As the territorial coincidence of mass production for mass consumption receded – as capitalism became recomposed in a New International Division of Labour – the KWS gave way to the 'competition state' (Hirsch 1997). In a world beset with breakneck globalization, the state's primary role turned from that of managing a national economy organized around a labour-capital accord to that of promoting its territory as a site, within an incipiently 'borderless world', for financialized capitalists searching for the best profit prospects. Turning away from material concessions to labour and other subaltern groups, the neoliberal state's mission was envisaged as 'protective' rather than compensatory. 'Protective democracy', as Neufield points out, enforces a stringent separation of the economic and the political, with the former protected from the latter and attuned only to the logic of the market while the state restricts its rule to 'allowing that logic to proceed without interference' (Neufield, 2001:102).

In this chapter I trace the shifts in accumulation, imperialism and hegemony that, since the 1980s, have attempted to restore conditions for robust 'growth' and a serviceable organization of consent, within a world order marked by declining American hegemony and rising geopolitical conflict, even as the ramifications of capitalism's ecological death drive became clearer and clearer. As we saw in Chapter 2, by the mid-1970s, the collapse of the long post-war boom mandated a paradigmatic change in the form of global capitalism. From the Fordism and 'embedded liberalism' of a US-led world order, the capitalist locomotive switched tracks to what Bonanno

(2000) calls Global post-Fordism (explained below). As the interrelated processes of financialization, globalization and neoliberalization took hold, a new world, arguably even more ecocidal than the Fordist regime, came into being.

The new paradigm for managing Fossil Capitalism's trifecta of accumulation, imperialism and hegemony brought with it several transformations. Not all of them were specific to the fossil fuel industry, but they all had implications for climate and for Fossil Capitalism as a way of life. I will first describe the main political-economic changes, followed by the cultural-political shifts in the post-Fordist era, which reached an inflection point with the Global Financial Crisis of 2008. Global capitalism has actually been on life-support in the years since that crisis, as the neoliberal zombie stumbles forward. The latter sections of this chapter reflect on this death drive and its ecocidal implications.

Political-economic Changes

If Fordism used deskilled assembly-line production to service mass consumption within a Northern-centred imperial mode of living, post-Fordism strove to recover the basis for sustained corporate profitability by incorporating computer-assisted, customized production for niche markets, in economies of scope instead of scale (Harvey 2006). In the North, largely unionized workforces were downsized, portending a decline in industrial unionism and thus a lower-wage economy. Within the new international division of labour, TNCs optimized profits by off-shoring mass production to the low-wage capitalist semi-periphery. On the back of Apple's iconic i-phones we are assured that the devices are 'designed in California'; however, the component parts are produced by workers in many countries (South and North) in companies under contract with Apple, with final assembly mainly in Shenzen, China. The watchword for post-Fordism has been 'flexibilization' – flexibility on a global scale – as 'mobile capital, free to colonize and commodify practically every sphere' shattered Fordism's relatively fixed boundaries (Bonanno 2000:313). In the process, production was dispersed to a plethora of companies controlled by transnational lead firms such as Apple. New 'flexible' structures also enabled transnational capital to access lucrative financial and tax arrangements offered by competition states seeking investment. We must not conflate flexiblization with a move away from monopoly capital's oligarchic power structure. Accompanying decentralized production was increasingly centralized control of finance, research and information within the transnational parent firms (Robinson 2004). Finally, 'global Post-Fordist "economic development" and free trade policies utilize[d] the State itself to enhance capital mobility, erode its own local, regional, and national regulatory instruments, and reduce labor's bargaining power and influence' (Bonanno 2000:313).

Indeed, across the global North the shape and form of the capitalist state also changed. The 'competition states' of neoliberalism, rejecting the redistributive role of nationally focussed Keynesian welfare states, have also been described as Schumpeterian workfare post-national regimes. Within such regimes international competitiveness is pursued through supply-side measures to fuel innovation in open economies, together with the subordination of social policy to economic objectives (attacking welfare rights and diminishing the 'social wage') (Jessop 2002:252).

In Chapter 1, I introduced the concept of the ecological debt which broadly speaking rises as the advanced capitalist zone depletes Southern ecosystems by withdrawing integral aspects of living systems while dumping by-products of extraction (including carbon emissions), as pollution. Capital's calculus, which is hegemonic, views debt quite differently. As they pursued capitalist development strategies in the 1960s and 1970s, many states in the South incurred enormous monetary debts, which, within neoliberal logic could only be repaid by adopting brutal austerity programmes. From the 1980s forward, the key institutions of global finance, the International Monetary Fund and World Bank, commenced a project to liberalize and integrate the global economy through 'structural adjustment' packages. Dubbed the 'Washington Consensus', these packages exemplified the new imperialism of accumulation by dispossession, as they

> regularly called for lowering or eliminating restrictions on capital mobility; the privatisation of public sector assets and industries; and wide-spread deregulation; in addition to restrictions on fiscal deficits. Presented as necessary interventions to bring supposedly 'backward' or protectionist economies into the modern globalised world while dealing with 'balance of payments' crises, the ultimate effect of these 'structural adjustments' was to open developing economies to (primarily) Northern private capital.
>
> *(Buller 2022:201)*

To earn the foreign exchange that covered interest payments on the debt, many Southern countries implemented policies that combined accelerating extractivism with austerity, degrading both ecosystems and the conditions of life for their citizens. In the global South, ecological imperialism and economic imperialism have combined to deliver a one-two punch to human communities and nonhuman nature. In contrast to capital's calculus, Adrienne Buller concludes, 'rather than financial debts owed by South to North, the world's wealthy economies owe an exorbitant human and ecological debt to the people and places that have produced our wealth and absorbed our waste' (2022:223).

In the era of Global post-Fordism the unprecedented globalization of capitalism brought with it a further internationalization of the imperial mode of living, within a continuing structure of imperialism. Alongside neoliberal policies to shift the balance of class power towards capital, which massively increased economic disparities both within countries and internationally (Milanovic 2024), the neoliberal era witnessed the global generalization of the imperial mode of living. Having 'universalized' across the global North in the era of Fordist embedded liberalism (Brand and Wissen 2021:89–99), in the era of Global post-Fordism the imperial mode became embedded (unevenly) in the global South – even as the basis for Fordist-style accumulation and state policy was undercut by capitalist globalization.

In the global North, as Brand and Wissen see things, capitalist globalization is now based on 'a new compromise between the elites and subalterns – in this case, especially the middle classes – a compromise that in essence contains a new deepening of the imperial mode of living. This compromise is tolerated and even largely approved of by many, thanks to the material opportunities for consumption that it provides' (2021:101). The deepening of the IML in the North has meant an escalating run on resources, North and South, including mountaintop removal coal mining in the United States and tar sands mining in Canada. It has meant the expansion of carbon-intensive industrial agriculture and associated land grabs and 'an ever-expanding and accelerating transport system', with airline passenger numbers increasing nearly tenfold between 1970 and 2015 (Brand and Wissen 2021:110, 112). These trends have exacerbated unequal ecological exchange; for instance, airline flights are accessible only to the relatively affluent, yet the intense carbon emissions from those flights add fuel to the fire for all humanity.

The IML's deepening in the North occurred as neoliberal post-Fordism sharpened economic disparities, excluding large segments of the working class from living the 'American Dream' of suburban automobility. Although its socio-material basis has been eroded by decades of neoliberal globalization, the IML remains a hegemonic standard embraced by many in the North – even if home ownership, for instance, is increasingly out of reach for most proletarians. In this sense, the 'imperial mode of living' is *increasingly no more than an aspiration*, for many working-class people in the global North. Yet that faint hope continues to have ideological power, particularly if one accepts the Thatcherite line that 'there is no alternative' to neoliberal capitalism. This clinging to the promised prosperity of a class compromise long ago abandoned by capital is one reason why Northern working class majorities are for the most part demobilized. But there are other compelling reasons, in particular the class politics of neoliberalism itself, which have disorganized and atomized working classes, leading to a pervasive resignation from politics (Chibber 2022:80).

Meanwhile, the imperial mode of living has been spreading selectively to affluent class fractions in the global South, deeply exacerbating the climate crisis and sharpening geopolitical tensions, as opportunities to externalize the impact of this mode of living diminish with its generalization. The UN estimates that the percentage of people living in North America and Europe (a proxy for the global North) will fall from 17 percent in 2023 to 14 percent in 2050, while world population increases from 8.05 billion to 9.71 billion (https://ourworldindata.org/region-population-2100). Widespread adoption of an unsustainable way of life by the majority world is an ecocidal prospect, underlining the need for a solution to the climate crisis that moves decisively beyond eco-imperialism and Fossil Capitalism. In the BRICS and other high-growth countries, the IML 'is becoming the dominant model of prosperity, even for those who have not yet been integrated into the imperial mode of living' (Brand and Wissen 2021:114). In adopting 'the "American way of life" with its individual transport, meat-heavy diet and consumer goods that rely strongly on natural resources' (Brand and Wissen 2021:116), vast numbers of middle-class Southerners are contributing to massive resource consumption, emissions and ecological degradation.

As we have seen, the IML has functioned within an eco-imperialist logic through which the North has been able to externalize, onto the South, the socio-ecological consequences of this way of life. But its universalization diminishes that ability, as more and more countries also externalize socio-ecological costs, 'thus competing with the global North both economically and ecologically' (Brand and Wissen 2021:132). The effect, as Brand and Wissen note, has been an intensification of global ecological crisis, 'increasing eco-imperial tensions among the countries of the global North and between them and the emerging powers of the global South' (Brand and Wissen 2021:133). The tensions, guaranteed to rise as the IML generalizes and climate breakdown accelerates, appear in various ways – from the rising volume of climate refugees, desperate for entry to Europe and the United States, to the struggles at COP meetings over which country is to reduce its carbon emissions, and by how much (ibid.).

The era of Global post-Fordism has also witnessed processes characterized by the eponymous terms McDonaldization and Walmartization. George Ritzer saw in McDonaldization an extension of Fordism's mass production/mass consumption accumulation regime into new realms, as low-wage, hyper-rationalized workplaces provide cheap, fast food to largely working-class consumers. As more and more retail businesses adopted the McDonald's franchise model, a good deal of social reproduction was repositioned within the 'cathedrals of consumption' that now dot urban landscapes around the world (Ritzer 2001:5). In this sense, the Fordist logic of mass production/mass consumption did not disappear in the post-Fordist era but, alongside post-Fordism, was itself globalized within the NIDL and Imperial Mode of Living. The related term, Walmartization, captures this element.

As corporations like Walmart tap into globalized commodity chains to move products *en mass* to working-class consumers, 'Walmartization has come to stand for the process that negatively impacts the growing armies of service workers, their communities, and the larger environment' (Lang and Klein 2015:205). In the post-Fordist era, this process, premised on very high rates of exploitation in the global South and vicious anti-union management strategies at Walmart and related enterprises, has delivered cheap consumer goods to workers in the global North. Even as their jobs have been off-shored and their wages have stagnated or fallen, they live the dream of affluence – jaded but still intact. Walmart has emphasized in its branding a strong commitment to environmental sustainability. But as Lang and Klein conclude, Walmart's sustainability paradox, embracing sustainability outside of its organization's supply chain but not inside its stores with respect to its workforce and their communities 'is not surprising given its hegemonic acceptance of the unsustainable idea that unbridled neoliberal consumer capitalism is somehow compatible with environmental health and social justice' (2015:206). McDonaldization and Walmartization have been integral to the generalization of the imperial mode of living. They have instantiated both economic and ecological unequal exchange, as low-paid workers in the global South subsidize mass consumption for workers in the global North undergoing their own immiseration.

Not surprisingly, with global generalization of the imperial mode of living and the elaboration of the new international division of labour the density and length of global commodity chains, including those carrying fossil fuels, took a quantum leap. Global commodity chains, linking extractive activities often on the capitalist periphery to manufacturing and consumption often in the core, have developed with innovations in transportation, from the 16th century onward, as the infrastructure for ecological unequal exchange (Ciccantell and Smith 2009).[1] Given that more than 80 percent of world trade is controlled by TNCs, 'these commodity chains can be seen as fastened at the center of the world economy, connecting production, located primarily in the global South, to final consumption and the financial coffers of monopolistic multinational firms, located primarily in the global North' (Suwandi et al. 2019:2). Alongside unequal ecological exchange, global commodity chains enable the enormous profits that transnational capital extracts from the Southern proletariat:

> Flexible, globalised production means that the most labour-intensive links in global commodity chains are located in the global South, where the reserve army of labour is larger, unit labour costs are lower, and rates of exploitation are thus correspondingly higher. The result is much higher profit margins for multinational corporations, with the additional value generated often credited to production in the center itself and with the overall process leading to the amassing of wealth in the center, via a kind of profit by expropriation.
>
> *(Suwandi et al. 2019:18)*

These chains comprise the 'global oil assemblage', part of which Watts (2021) has tracked from well-heads in Nigerian villages. But the signature of fossil capital is written on commodity chains that may seem far removed from extractive capitalism. Kate Crawford's study of the artificial intelligence industry reveals how five of the world's largest tech corporations, all based in San Francisco, rely on materials extracted elsewhere for their super-profits. Although AI seems to lift us out of the material world, into the cloud, in fact 'the cloud', the backbone of the internet and artificial intelligence industry, is 'made of rocks and lithium brine and crude oil' (2021:31).

Amid these globalizing developments, in the neoliberal era natural gas gained importance in the fossil energy mix, as the development of energy-intensive technologies for liquefying gas and transporting it overseas lowered production costs. Liquified Natural Gas (LNG) became a major fossil fuel, accompanied by commodity chains and fixed-capital facilities to liquefy, transport, regassify and transport to final burning. In 1970, natural gas accounted for 14 percent of global primary energy consumption (compared to 39.9 percent for petroleum and 25.8 percent for coal). By 2022 natural gas supplied 22.0 of primary energy, petroleum 29.6 percent and coal 25.1 percent (Ritchie and Rosado 2024).

These globalizing aspects of accumulation also brought a tendential transnationalization of class itself, including the capitalist class and its organic intellectuals. The transnational network of interlocking directorates among the largest corporations expanded, as did policy spaces like the World Economic Forum, Trilateral Commission and the World Business Council on Sustainable Development (the last being corporate capital's attempt to manage capital's 'second contradiction', Carroll 2010; Carroll and Sapinski 2016). All of them mechanisms for neoliberal global governance, these formations brought together economic and political elites but excluded labour and popular voices. Robinson (2004) exaggerated the extent to which a transnational capitalist class has displaced nationally based capitalist classes (Carroll 2018). Yet there is no doubt that these developments, in step with the globalization of capital itself, enhanced the ability of the top tier of capitalists and their organic intellectuals to develop a 'full-spectrum' neoliberalism – an array of 'short-, middle- and long-term initiatives based in "the immutable solidity of the market"' (Mirowski 2013:342).

That 'immutable solidity' ramified throughout the imperialist order. From the 1980s on, as the Washington Consensus defeated economic nationalism in the global South, subject states were integrated into the debt trap of neoliberal imperialism. IMF structural adjustment policies not only mandated 'free' markets, with minimal state regulation, along with privatization, trade liberalization and limited state budget deficits (Peet 2009). They also ensured that Northern banks would see their loans repaid. Further, they exterminated the threat of economic nationalism that post-colonial states posed in the 1950s, 1960s and 1970s.

> Under neoliberal reforms, third-world ecological resources, and the profits from them, were once again made accessible to the demands of capital accumulation, as lingering obstructions enacted during the developmentalist era were removed, one after the other. Within a context of economic and political crisis, the economic nationalist counter-movement, born from the contradictions engendered by the expansion of the world capitalist system, was replaced by the neoliberal counterrevolutionary movement, the current phase of ecological imperialism.
>
> *(Frame 2022:521)*

Jennifer Rice and her colleagues (2022) have characterized the new regime of ecological imperialism as 'climate apartheid', evidenced by 'uneven vulnerabilities to the climate crisis, as well as inequitable implementation of climate-oriented infrastructures, policies, and programs. These efforts often secure privileged populations while harming, excluding, and criminalizing populations whose lives have been made precarious by climate change' (2022:625). A 'co-produced system of privilege and precarity', climate apartheid results from both the material effects of climate change and the responses to the climate crisis. The combination

> coproduces two populations: the climate privileged, those with the wealth and subject positions to insulate themselves from the greatest threats of climate change, and perhaps even profit from them; and the climate precarious, those whose social status and lack of access to safe and resilient infrastructures may facilitate or exacerbate their vulnerability, harm, or displacement.
>
> *(Rice et al. 2022:627)*

Climate apartheid is co-produced in three ways:

> (1) through uneven vulnerability to climate-related hazards and unequal access to resources and ecologies necessary for a happy and healthy life, (2) through climate-related interventions in the built environment (namely housing and infrastructure) that benefit some more than others, and (3) through increasingly exclusive conceptualizations of citizenship, especially as they relate to migration and securitization under climate change.
>
> *(Rice et al. 2022:629)*

The last of these is particularly worrisome, as

> systems of privilege are reinforced through hardened borders and new ideas of citizenship under climate change, while precarity is made worse through increasing exclusivity of mobility and migration. ... Like South African

> apartheid, climate apartheid is the culmination of a systematic, historical project of dehumanization, dispossession, violence, and exploitation.
>
> *(Rice et al. 2022:632)*

Juxtaposed with the generalization of the imperial mode of living, climate apartheid points us to the deeply contradictory political ecology of contemporary global capitalism, as increasing precarity, both South and North, excludes most people from the IML, even as it is celebrated by some as a universal capitalist utopia. Indeed, climate apartheid, ecological imperialism and full-spectrum neoliberalism have thinned transnational capitalism's historical bloc, and weakened the basis for hegemony. The exclusionary character of transnational neoliberalism brought class compromise to a decisive end, as large segments of the populace, both North and South were excluded from the material and symbolic concessions that had been the basis of the accord (Cox 1987). The current world order, as Eve Croeser points out, 'is categorised as non-hegemonic because the project of capitalist globalisation that prevails cannot make the concessions that would elicit the widespread support from global civil society that, according to Cox and other neo-Gramscian analysts, is the prerequisite of hegemony' (2020:164). The clearest indicator of this failing hegemony has been the rise of mass movements, in national theatres and globally, opposing corporate globalization in the 1980s and 1990s, and in more recent years, opposing imperialist wars, financialized capitalism (Occupy Wall Street) and ecocide.

The weakening of hegemony in the global capitalist order has closely tracked the decline of US hegemony, which provided leadership within that order across the post-war era. Already in the late 1970s Japan began to overtake the United States as an industrial competitor, followed by western Europe in the 1990s. The United States doubled down, weakening its rivals.[2] The dissolution of the Soviet Union in 1991 marked the coming of a 'unipolar world' to be governed by the United States. Yet American hegemony was already a fading star in the firmament of global capitalism; hence, to shore up its dominance the United States turned increasingly to militarism in place of consensual power. In the 1970s, the defeat of the United States by communist insurgents in Viet Nam had already foreshadowed the limits of American hard power. Its embarrassing exits from wars it provoked in Afghanistan (2001–21) and Iraq (2003–10) underlined those limits further. Continuing a legacy of oil imperialism, the struggle to control petroleum resources in west Asia sharpened, with US-led wars on Libya (2011) and Syria (2014).

By 2002, in the wake of the September 11, 2001 attacks and a few months before the US invasion of Iraq, leading sociologist Immanuel Wallerstein declared Pax Americana finished, stating that 'the eagle has crash landed.' To Wallerstein, the long season of American economic and military failures revealed 'the limits of American supremacy. Will the United States learn to fade quietly, or will U.S. conservatives resist and thereby transform a gradual decline into a rapid and dangerous fall?' he asked (Wallerstein 2002:60). As global capitalism subsequently descended into global disorder and geopolitical rivalry, driven partly by the assertion of American hard power, it became clear that the latter option would be the road taken. The disorder did nothing to slow the accumulation of fossil capital and atmospheric carbon (warfare itself being a major contributor to the latter), and the emergent regime of neoliberal financialization seemed to put climate solutions further and further out of reach.

Ideological Shifts

Accompanying the massive shifts in geopolitical economy, the coming of Global post-Fordism heralded changes to the shape and form of ideological hegemony as it lives within us. I will try to summarize these developments in five points.

- The shift from a Fordism favouring industrial capital to a Global post-Fordism built around financialization heightened the *market fetishism* that consumer capitalism had already encouraged. The concern with planning and macroeconomic industrial strategies was minimized in favour of a quasi-religious faith in the rationality and efficiency of markets, expressed by political leadership and through mainstream media.
- Neoliberalism promoted the *possessive individual*, that is, individual choice and the 'uncaring' part of human psychology 'over the reality of our interdependent relationships with each other, and our connections with external realities like climate change.' In enticing individuals to consume without any sense of collective responsibility, 'the broader neoliberal mindset has a natural overlap with the mindset of the Fossil Empire' (Betts 2021:210).
- Within the neoliberal world view, the treadmill of production could beneficially speed up, intensifying consumption and avoiding any reduction in working hours – thus promoting 'endless accumulation and growth' (Buch-Hansen and Nielsen 2023:356). In 1930, as the Great Depression was settling in, economist John Maynard Keynes predicted that productivity improvements would eventuate in a 15-hour workweek (Krook 2017). Yet 'under the hegemony of neoliberal ideas, the goal to reduce work time disappeared from the political agenda in the 1990s and later it was altogether forgotten' (Buch-Hansen and Nielsen 2023:356). Meanwhile, the levels of work intensity and consumption relentlessly increased, along with carbon emissions.
- In this Global post-Fordist era, with the *intensification* of consumer capitalism, capital accelerated its colonization of lifeworlds, aligning identities

with symbolic branding, from Apple i-phones to Zoya lipstick. In the global North, Fredric Jameson famously characterized postmodernism as 'the cultural logic of late capitalism' (1984). As David Tetzlaff (1991) pointed out, the ideological import of postmodernism lay in its fragmenting impact on social consciousness, arising in part from the post-Fordist shift from mass markets to niche markets, activating many diverse identities associated with many brands and consumer choices. Postmodern cultural fragmentation plays on both the proliferation of stylized subcultures detached from each other and the fascination with mediatized spectacle which 'effectively abolishes any practical sense of the future' (Jameson, 1984:85), and therefore any interest in a transformative political project (Tetzlaff, 1991:29–30). In the neoliberal era such fragmentation became aligned with a divide-and-conquer hegemonic strategy in which consent is won without a clear ideological consensus. I am writing this a day after a record 123.7 million Americans tuned into Super Bowl 58 (many of them to see mega-star Taylor Swift; Porter 2024) – roughly the same number of Americans who vote in presidential elections. In the more postmodernized societies like the United States, such spectacles bring the imagined community together under the mantle of nationalism, without implying any political consensus that goes beyond faith in consumer capitalism itself. '*In the spectacle*,' as Kevin Gotham and Daniel Krier emphasize, '*media and consumer society replace lived experience, the passive gaze of images supplants active social participation, and new forms of alienation induce social atomization at a more abstract level than in previous societies*' (2008:155, emphasis in original).
- Indeed, a telling symptom of the intensification of consumer capitalism and the hollowing-out of liberal democracy was the gradual shift in core identity from citizens to consumers. As we saw in Chapter 3, this began already during post-war Fordism. It was further amplified under neoliberalism in the form of a *hostile privatism* centred in the suburban experience. Again, the US case is illuminating. 'Before long, much of popular common sense simply forgot about the public basis of a privatized existence' (Huber 2013:94). In truth, 'one of the greatest violences of the neoliberal era was the closure of the political imagination' (Smith 2010:56), as market fetishism compressed 'the horizon for alternatives onto the head of a pin, rendering as natural, inevitable and inescapable a world dominated by market dynamics and governed by the price mechanism' (Buller 2022:97). 'Revolutionary possibility was generally confused with utopianism ... and revolution was collapsed into a caricature of inevitable failure' (Smith 2010:56).

These ideological shifts guaranteed that, from a hegemonic perspective, the developing climate crisis would be addressed as 'matters of individual consumption choices, cost–benefit analyzes and market-based solutions.' Carbon-trading schemes, which turn the climate crisis 'into an investment opportunity by creating markets in which emissions certificates are traded and speculative profits can be made' became exemplary policies (Buch-Hansen and Nielsen 2023:356). As we shall see in Chapter 4, these schemes have proven ineffective given the scale and urgency of the crisis.

As Fordism's contradictions deepened in the late 1960s, a New Left, loosely enveloping several emergent political currents (so-called New Social Movements (NSMs)) emerged, including modern environmentalism and, by the 1970s, Green parties across the global North. NSMs comprised 'a new battleground in the struggle for a new hegemony' (Hirsch 2011:51). From the 1970s onward, ecological issues became contested terrain in public discourse, with capital pushing a jobs-vs.-environment discourse to divide and rule and environmentalists typically pushing narrow agendas of conservation, inured to the concerns of most proletarians (Foster 1993). As the contradictions in neoliberal globalization deepened, a counter-movement emerged in the 1990s, bringing together leftists, labour activists, environmentalists, feminists and anti-colonialists under the banner of alter-globalization in Europe and anti-corporate globalization in North America. Amid these contradictions, a diverse collection of NSMs, some with roots in the late 1960s, was consolidated, consistent with O'Connor's interpretation in his foundational article on capitalism's second contradiction, that such struggles over production conditions are 'not less but more than class issues' (O'Connor 1988:34), and crucial in opposing the second contradiction (cf. Lebowitz 2020; Fraser 2022).

A dialectical perspective on accumulation, imperialism and hegemony recognizes that ruling-class projects such as the Fordist-KWS or neoliberal post-Fordism always face resistance from below, even if that resistance is largely ineffectual. Just as the 19th-century workers' movement provided the impetus and context for the emergence of historical materialism, other liberatory movements in the past century and a half have inspired emergent streams of critical thought, including feminism, anti-colonialism and environmentalism. Clearly, any coalition of forces opposing Fossil Capitalism needs to knit them together, into an incipient historical bloc. This is a key challenge in the climate crisis, to be addressed in depth in Chapters 5 and 6, after we have dispensed in Chapter 4 with false solutions that come from on high.

Global Slump and the Neoliberal Zombie

The various political-economic and ideological changes I have recounted above characterized the era of neoliberal globalization and deepening climate crisis, up to our current circumstances. However, for neoliberalism the global financial crisis of 2008 was an inflection point, after which the project struggled to regain viability. The crisis – cascading from massive asset devaluation in the US mortgage market to insolvent investment banks, crashing stock markets and, by late 2008, a 'frozen' international financial system threatening worldwide depression – was greeted by many as a sign of neoliberalism's demise. Yet, in the aftermath of the meltdown, it became clear that the crisis 'actually granted neoliberalism a second life' (Carroll and Sapinski 2016:32; cf. Mirowski 2013; Bruff 2014). But that second life has amounted to little more than a zombified existence.

Until the 2008 financial crisis, the project of neoliberal globalization had been rolling out 'as planned', though not without various wrinkles. Neoliberal policy did revive accumulation for a quarter of a century (roughly 1983 to 2008), but ballooning financialization rendered global capitalism more unstable (McNally 2011).[3] Given capitalism's internal contradictions, it was neoliberalism's success that set the table for its collapse in the 2008 financial crisis. Post-Fordist flexibilization, new information technologies, deregulation, austerity, attacks on organized labour and accumulation by dispossession did stoke accumulation, but these policies also immiserated large segments of the proletariat, driving many into unsustainable debt often tied to home mortgages. As the neoliberal boom crested, capital poured into speculative financial markets offering high yields on the promise of endless good times. The neoliberal boom ended dramatically with a crash comparable in scale to the 1929 financial panic that triggered the Great Depression. Massive coordinated intervention by leading capitalist states saved capitalism from itself, largely through bank bailouts. The bailouts created enormous public debt, covered via more neoliberal austerity. Yet with the exception of China, which departs from neoliberalism with its extensive economic planning and regulations on capital, the ensuing years have not seen a return to robust accumulation. Instead, according to Chris Harman, '21st century capitalism as a whole is a zombie system, seemingly dead when it comes to achieving human goals and responding to human feelings, but capable of sudden spurts of activity that cause chaos all around' (2009:11).

To be sure, global capitalism has descended into another organic crisis, as no viable alternative to neoliberalism has emerged from either the capitalist side or a disorganized left. Gramsci's description of organic crisis, offered during the Great Depression and quoted earlier, is worth repeating: 'the old is dying and the new cannot be born, in this interregnum, a great variety of morbid symptoms appear' (Gramsci 1971:276) – chief among them the rise of the far right and neo-fascism.

The crisis we are living is unique in human history. Earlier, I explained how capitalism is a crisis-dependent form of economy and society. Accumulation moves in cyclical waves, with a rhythm attuned to changing prospects for realizing profit. In crises, the downswing becomes a catalyst for creation of new conditions for accumulation. The ongoing crisis of neoliberalism fits this description. In principle, it is feasible for such conditions to be formed.

However, capitalism does not float freely in its own medium; it is immersed in nature. Until capitalism reached global scale in the late 20th century, ecological crises generated by the treadmill of production were local – a polluted river here, a clear-cut forest there. But since the Great Acceleration following World War 2, capitalism, ever an irritant, has become 'the enemy of nature' (Kovel 2007). Today's organic crisis is both a deep crisis of capitalism and a global ecological crisis. The latter, however, follows no cyclical pattern. Instead, it portends *the cumulative breakdown of the material conditions for the living systems upon which our existence depends*, that is, ecocide. 'The ravages of the climate crisis are emerging amidst considerable upheaval in capitalist hegemony' – and vice versa. 'The patchwork deterioration of neoliberalism amidst the social and ecological crises exacerbated by that very system is generating an organic crisis, wherein capital is failing to deliver the socio-ecological goods, and suffering a genuine crisis of legitimation' (Surprise 2024:448–9).

Reflecting the grim future likely to stem from business-as-usual, Oil Change International has analyzed the carbon footprint of planned oil and gas extraction, from 2023 to 2050 – a projected global total of 195.5 gigatonnes of CO_2. The countries leading the pack, the 'Planet Wrecker 20' are set to contribute 88.4 percent of the additional carbon pollution. 'The United States is Planet Wrecker In Chief', accounting for 37.1 percent of planned global oil and gas expansion through 2050, followed by Canada (9.5 percent) and Russia (8.9 percent) (Loualalen and Trout 2023:4).[4] Oil Change International concludes, 'If these [20] countries proceed with their new extraction, committed carbon pollution will be 190 percent over the 1.5°C budget' (ibid.), locking the planet into a 'hothouse earth' scenario of catastrophic climate breakdown.

Meanwhile, 'the post-Cold War order, allegedly built upon the "golden arches of peace," is on the brink of another historical phase of escalating international conflicts, consisting of a combined geoeconomic and geostrategic rivalry, between competing capitalist states and blocs in the context of an increasingly multipolar world order' (Hosseini and Gills 2023:6). The toll in human death and suffering from major wars underway as I write these lines is horrific. But militarism also contributes mightily to the climate crisis. The world's military forces have a carbon footprint conservatively estimated to be greater than the entire continent of Africa (Schlanger 2024) – leaving aside 'GHG emissions arising from the impacts of warfighting, such as fires, other damage to infrastructure and ecosystems, post-conflict reconstruction, and health-care for survivors' (Parkinson and Cottrell 2022:8). Actual kinetic warfare adds not only death and destruction but vast carbon emissions to this 'base budget.'[5]

Conclusion: From Death-drive to a Liveable World?

If climate change is, as Al Gore famously said, an inconvenient truth, other worries of our era include the unfortunate coincidence of that truth with neoliberalism (Klein 2014) and, most recently the heightened militarism in a collapsing US-led international order, just as the need for peaceful and serious climate diplomacy is most urgent. The failure of such diplomacy at annual COP meetings, infested as they are with fossil capitalists and their shills, has added to the worries.

We have seen in this chapter how, by the second quarter of the 21st century, neoliberal globalization had spent itself (with no alternative in sight) while severely exacerbating the climate crisis. In the process, the means for solving the organic crisis of capital and Earth had weakened to the point of being indiscernible. In the chapter that follows I will chart the dubious measures that have been proposed, within the regime of neoliberal global post-Fordism, to avoid climate catastrophe. A pathway towards a liveable world will require a decisive break from that regime, and, indeed, from capitalism as a way of life. In the final two chapters of this book, I take up those possibilities.

Notes

1 Ciccantell and Smith (2009:373) report that in 1400 state-of-the-art shipping technology capacitated approximately 400 tons of cargo, but that by 2008 the largest bulk-shipping vessels could carry 400,000+ tons. Ortiz-Ospina et al. (2018) report that between 1930 and 2005 the costs of shipping by sea fell by 78% while the cost of passenger air transport fell by 98%.

2 The Plaza Accord of 1985, requiring Japan to revalue the Yen upward, set Japan on a course towards economic stagnation, from which it has not recovered (Beeson and Broome 2010). In the 1990s, US-led NATO actions in the former Yugoslavia kept a Europe (then in the process of integrating through the EU) relatively divided and enabled American power to expand by enlarging NATO's mandate into eastern Europe and beyond, collaterally securing oil resources located in former Soviet republics (Cox 2020).

3 Financialization has continued unabated since the 2008 crisis. 'In 2010, global financial assets had a total value of 211 trillion USD. This corresponds to 356 per cent of the same year's global GDP, which was 66 trillion USD. By 2020, the total value of global financial assets had risen to 418.3 trillion USD, while global GDP amounted to 85 trillion USD, which meant that global financial assets were worth 492 per cent of global GDP' Klein 2014:4 (Klein 2024:4).

4 In contrast, China, which since 2018 has embraced an 'Ecological Civilization' in its constitution and its national development strategy (Wei et al. 2021) is projected to account for 4.6% of projected carbon emissions from new fossil-fuel projects.

5 The first year of the Ukraine-Russia war in 2022 is estimated to have had a carbon footprint equivalent to that of Belgium (Neimark 2023).

References

Beeson, Mark, and André Broome. 2010. "Hegemonic Instability and East Asia: Contradictions, Crises and US Power." *Globalizations* 7(4):507–23. doi: 10.1080/14747731003798500

Betts, Alan K. 2021. "Climate Change and Society." *AIMS Geosciences* 7(2):194–218. doi: 10.3934/geosci.2021012

Bonanno, Alessandro. 2000. "The Crisis of Representation: The Limits of Liberal Democracy in the Global Era." *Journal of Rural Studies* 16(3):305–23. doi: 10.1016/S0743-0167(99)00064-9

Brand, Ulrich, and Markus Wissen. 2021. *The Imperial Mode of Living*. London: Verso.

Bruff, Ian. 2014. "The Rise of Authoritarian Neoliberalism." *Rethinking Marxism* 26(1):113–29. doi: 10.1080/08935696.2013.843250

Buch-Hansen, Hubert, and Peter Nielsen. 2023. "Critical Realism, the Climate Crisis and (de)Growth." *Journal of Critical Realism* 22(3):347–63. doi: 10.1080/14767430.2023.2217050

Buller, Adrienne. 2022. *The Value of a Whale: On the Illusions of Green Capitalism*. Manchester: Manchester University Press.

Cahill, Damien. 2012. "The Embedded Neoliberal Economy." Pp. 110–27 in *Neoliberalism: Beyond the Free Market*, edited by D. Cahill, L. Edwards, and F. Stilwell. Cheltenham, UK: Edward Elgar.

Carchedi, Guglielmo, and Michael Roberts, eds. 2018. *World in Crisis*. Chicago: Haymarket Books.

Carroll, William K. 2006. "Hegemony, Counter-Hegemony, Anti-Hegemony." *Socialist Studies/Études Socialistes* 2(2). doi: 10.18740/S44G7K

Carroll, William K. 2010. *The Making of a Transnational Capitalist Class*. London: Zed Books.

Carroll, William K. 2018. "Rethinking the Transnational Capitalist Class." *Alternate Routes* 29:188–206.

Carroll, William K., and Jean Philippe Sapinski. 2016. "Neoliberalism and the Transnational Capitalist Class." Pp. 25–35 in *The Handbook of Neoliberalism*, edited by K. Birch, J. MacLeavy, and S. Springer. London: Routledge.

Chibber, Vivek. 2022. *The Class Matrix*. Cambridge, MA: Harvard University Press.

Ciccantell, P., and D. A. Smith. 2009. "Rethinking Global Commodity Chains: Integrating Extraction, Transport, and Manufacturing." *International Journal of Comparative Sociology* 50(3–4):361–84. doi: https://doi.org/10.1177/0020715209105146

Cox, Robert W. 1987. *Production, Power, and World Order: Social Forces in the Making of History*. New York: Columbia University Press.

Cox, Ronald W. 2020. "US Imperialism and the Disintegration of Yugoslavia." Pp. 1–17 in *The Palgrave Encyclopedia of Imperialism and Anti-Imperialism*, edited by I. Ness and Z. Cope. Cham: Springer International Publishing.

Crawford, Kate. 2021. "Earth." Pp. 23–51 in Kate Crawford, *Atlas of AI*. New Haven: Yale University Press.

Croeser, Eve. 2020. *Ecosocialism and Climate Justice: An Ecological Neo-Gramscian Analysis*. London: Routledge.

Dietz, Simon, James Rising, Thomas Stoerk, and Gernot Wagner. 2021. "Economic Impacts of Tipping Points in the Climate System." *Proceedings of the National Academy of Sciences* 118(34):e2103081118. doi: 10.1073/pnas.2103081118

Foster, John Bellamy. 1993. "The Limits of Environmentalism without Class: Lessons from the Ancient Forest Struggle of the Pacific Northwest*." *Capitalism Nature Socialism* 4(1):11–41. doi: 10.1080/10455759309358529

Frame, Mariko L. 2022. "Ecological Imperialism: A World-Systems Approach." *The American Journal of Economics and Sociology* 81(3):503–34. doi: 10.1111/ajes.12472

Fraser, Nancy. 2022. *Cannibal Capitalism: How Our System Is Devouring Democracy, Care, and the Planet – and What We Can Do about It*. London; New York: Verso.

Gamble, Andrew. 1994. *The Free Economy and the Strong State*. Second. Leiden: Springer.

Gotham, Kevin Fox, and Daniel A. Krier. 2008. "From the Culture Industry to the Society of the Spectacle: Critical Theory and the Situationist International." Pp. 155–92 in *Current Perspectives in Social Theory*. Vol. 25, edited by Harry F. Dahms. Bingley: Emerald Publishing Limited.

Gramsci, Antonio. 1971. *Selections from the Prison Notebooks*. edited by Q. Hoare and G. Nowell-Smith. New York: Lawrence & Wishart.

Harman, Chris. 2009. *Zombie Capitalism, Global Crisis and the Relevance of Marx*. Chicago: Haymarket Books.

Harvey, David. 2006. *The Condition of Postmodernity: An Enquiry into the Origins of Cultural Change*. Cambridge, MA: Blackwell.

Harvey, David. 2005a. *A Brief History of Neoliberalism*. New York: Oxford University Press.

Harvey, David. 2005b. *The New Imperialism*. New York: Oxford University Press.

Hirsch, Joachim. 1997. "Globalisation of Capital, Nation-States and Democracy." *Studies in Political Economy* (54):39–58.

Hirsch, Joachim. 2011. "The Crisis of Fordism, Transformations of the 'Keynesian' Security State." Pp. 43–57 in *Research in Social Movements, Conflicts and Change*, *Research in Social Movements, Conflicts and Change*, edited by P. G. Coy. Bingley: Emerald.

Hosseini, S. A., and Barry K. Gills. 2023. *Capital Redefined: A Commonist Value Theory for Liberating Life*. 1st ed. London: Routledge.

Huber, Matthew T. 2013. *Lifeblood: Oil, Freedom, and the Forces of Capital*. Minneapolis: University of Minnesota Press.

Intan Suwandi, R. Jamil Jonna, and John Bellamy Foster. 2019. "Global Commodity Chains and the New Imperialism." *Monthly Review* 70(10):1–24.

Jameson, Fredric. 1984. "Postmodernism, or The Cultural Logic of Late Capitalism." *New Left Review* 1(146):53–92.

Jessop, Bob. 2002. *The Future of the Capitalist State*. Cambridge: Polity Press.

Klein, Dieter. 2024. *Within and Beyond Capitalism: A Twofold Transformation*. London: Routledge.

Klein, Naomi. 2014. *This Changes Everything: Capitalism vs. the Climate*. Toronto: Alfred A. Knopf Canada.

Kovel, Joel. 2007. *Enemy of Nature*. London: Zed Books.

Krook, Joshua. 2017. "Whatever Happened to the 15-Hour Workweek?" *The Conversation*. Retrieved August 28, 2024. (https://theconversation.com/whatever-happened-to-the-15-hour-workweek-84781).

Lang, Steve, and Lloyd Klein. 2015. "Walmart's Sustainability Initiative: Greening Capitalism as a Form of Corporate Irresponsibility." Pp. 197–207 in *The Routledge International Handbook of the Crimes of the Powerful*, edited by Gregg Barak. London: Routledge.

Lebowitz, Michael A. 2020. *Between Capitalism and Community*. New York: Monthly Review Press.

Loualalen, Romain and Kelly Trout. 2023. "Planet Wreckers: How 20 Countries' Oil and Gas Extraction Plans Risk Locking in Climate Chaos." *Oil Change International*. Retrieved Aug 28, 2024. (https://coilink.org/20.500.12592/js1xqk).

McNally, David. 2011. *Global Slump*. Oakland, CA: PM Press.
Milanovic, Branko. 2024. "The Three Eras of Global Inequality, 1820–2020 with the Focus on the Past Thirty Years." *World Development* 177:106516. doi: 10.1016/j.worlddev.2023.106516
Mirowski, Philip. 2013. *Never Let a Serious Crisis Go to Waste: How Neoliberalism Survived the Financial Meltdown*. London; New York: Verso.
Neimark, Benjamin. 2023. "How to Assess the Carbon Footprint of a War." *The Conversation*. Retrieved February 14, 2024 (https://theconversation.com/how-to-assess-the-carbon-footprint-of-a-war-215575).
Neufeld, M. 2001. "Theorizing Globalisation: Towards the Politics of Resistance – A Neo- Gramscian Response to Mathias Albert." *Global Society* 15:93–106.
Nikiforuk, Andrew. 2024. "A New Report Maps Canada's 'Daunting' Path to Net-Zero Carbon." *The Tyee*. Retrieved February 8, 2024 (https://thetyee.ca/Analysis/2024/02/08/Canada-Daunting-Path-Net-Zero-Carbon/).
O'Connor, James. 1988. "Capitalism, Nature, Socialism a Theoretical Introduction." *Capitalism Nature Socialism* 1(1):11–38. doi: 10.1080/10455758809358356
Ortiz-Ospina, Esteban, Diana Beltekian, and Max Roser. 2018. *Trade and Globalization. Our World in Data*. Retrieved August 28, 2024 (https://ourworldindata.org/trade-and-globalization).
Parkinson, Stuart, and Linsey Cottrell. 2022. *Estimating the Military's Global Greenhouse Gas Emissions*. London: Scientists for Global Responsibility and Conflict and Environment Observatory.
Peet, Richard. 2009. *Unholy Trinity: The IMF, World Bank, and WTO*. 2nd edition. London: Zed Books.
Porter, Rick. 2024. "TV Ratings: Super Bowl LVIII Sets All-Time Record With 123M Viewers." *Hollywood Reporter*. Retrieved (https://www.msn.com/en-us/tv/news/tv-ratings-super-bowl-lviii-sets-all-time-record-with-123m-viewers/ar-BB1ibHMf).
Rice, Jennifer L., Joshua Long, and Anthony Levenda. 2022. "Against Climate Apartheid: Confronting the Persistent Legacies of Expendability for Climate Justice." *Environment and Planning E: Nature and Space* 5(2):625–45. doi: 10.1177/2514848621999286
Ritchie, Hannah, and Pablo Rosado. 2024. "Energy Mix." *Our World in Data*. Retrieved August 28, 2024 (https://ourworldindata.org/energy-mix).
Ritzer, George. 2001. *Explorations in the Sociology of Consumption: Fast Food, Credit Cards and Casinos*. London: Sage.
Robinson, William. 2004. *A Theory of Global Capitalism*. Baltimore: Johns Hopkins University Press.
Schlanger, Zoë. 2024. "Military Emissions Are Too Big to Keep Ignoring." *The Atlantic* (Jan. 17). Retrieved September 13, 2024 (https://www.theatlantic.com/science/archive/2024/01/military-emissions-climate-cop28/677151/).
Smith, Neil. 2010. "The Revolutionary Imperative." *Antipode* 41(s1):50–65. doi: 10.1111/j.1467-8330.2009.00716.x
Surprise, Kevin. 2024. "Beyond Ecocidal Capitalism: Climate Crisis and Climate Justice." Pp. 448–68 in *The Elgar Companion to Antonio Gramsci*, edited by W. K. Carroll. Cheltenham, UK: Edward Elgar.
Teeple, Gary. 2008. *Globalization and the Decline of Social Reform: Into the Twenty-First Century*. 2nd ed. Aurora, Ont., Amherst, NY: Garamond Press : Humanity Books.

Tetzlaff, David. 1991. "Divide and Conquer: Popular Culture and Social Control in Late Capitalism." *Media, Culture & Society* 13(1):9–33. doi: 10.1177/016344391013001002

United Nations. n.d. "What is the Kyoto Protocol?" *United Nations Framework Convention on Climate Change*. Retrieved September 13, 2024 (https://unfccc.int/kyoto_protocol).

Wallerstein, Immanuel. 2002. "The Eagle Has Crash Landed." *Foreign Policy* 131:60–68.

Watts, Michael John. 2021. "Space, Scale, and the Global Oil Assemblage: Commodity Frontiers in Resource Peripheries." Pp. 145–82 in *Resource Peripheries in the Global Economy*, edited by F. Irarrázaval and M. Arias-Loyola. Cham: Springer International Publishing.

Wei, Fuwen, Shuhong Cui, Ning Liu, Jiang Chang, Xiaoge Ping, Tianxiao Ma, Jing Xu, Ronald R. Swaisgood, and Harvey Locke. 2021. "Ecological Civilization: China's Effort to Build a Shared Future for All Life on Earth." *National Science Review* 8(7). https://doi.org/10.1093/nsr/nwaa279

PART II

Towards a Liveable World

4

THE FALSE SOLUTIONS OF CLIMATE CAPITALISM

Part I of this book provided the theoretical tools for understanding Fossil Capitalism and climate crisis (Chapter 1), followed by an analysis of how, since World War 2, an increasingly globalized capitalism powered by carbon energy has accelerated CO_2 emissions, placing humanity now at risk of ecocide (Chapters 2 and 3). Within a historical materialist perspective, our guiding theoretical framework has featured the trifecta of capital accumulation, imperialism and hegemony as core forms of power within late capitalism. This combination highlights the socio-ecological dynamics of capitalism as a class-based form of economy demanding endless growth within a structure of imperialism that reproduces vast North-South inequities, and the hegemonic projects that have legitimated Fossil Capitalism even as subaltern forces have resisted from below.

In Part II, the focus shifts to the contemporary scene, and the difficult task of avoiding ecocide, given the enormous physical inertia now driving climate breakdown and the formidable socio-political obstructions to transformative change. Within the trifecta of power that constitutes a transnational hegemonic bloc supporting capitalism as a way of life (not to be reified as a thing, but understood as a configuration of socio-ecological relations), solutions to the climate crisis must, fundamentally, reproduce that way of life. In this chapter I review the solutions emanating from the organic intelligentsia of the ruling class, each of which protects what Gramsci called 'the decisive nucleus of economic activity' (Gramsci 1971:161) from which the power of capital radiates. In subsequent chapters I take up approaches that recognize the urgent need to break from capitalism and imperialism – to refuse ecocide along with the way of life that is leading us to the brink.

DOI: 10.4324/9781003412922-7

The Illuminating Case of a Climate Laggard/leader

If results from a google search I just did (on 2 April 2024) were a reflection of reality beyond the Internet, Canada's political and economic leadership could bask in well-deserved glory. 'Canada climate leader' returns 662,000,000 urls; 'Canada climate laggard' returns 547,000 urls. The comparison is obviously inexact and a little unfair, since 'leader(ship)' is a featured term in state policy discourse, echoed by mainstream media and also deployed by climate activists, as they call for real leadership; while 'laggard' carries such negative connotations that only fierce critics of Canadian policy invoke it. Even so, we face a paradox, a yawning gap between reputation and reality. The Canadian case is illuminating, not as an exception but as an exemplar of the *regime of obstruction* found throughout the world of Fossil Capitalism. My first task in this chapter is to unpack how that exemplary regime has operated, drawing from research that a team I co-led undertook between 2015 and 2022, under the aegis of the Corporate Mapping Project (Carroll, 2021, 2022).

We saw in Chapter 3 that, as far as future plans go, Canada, a country one-tenth the size of the United States in population, ranks second to the United States globally in approved projects that will dump more carbon into the atmosphere, in the middle of a climate emergency. In fact, in the years since Prime Minister Justin Trudeau declared at the 2015 Paris COP meetings, 'Canada is back,' Canada's record in reducing emissions has been pathetic. Trudeau ascended to power after the 2015 election, replacing a Conservative government led by Stephen Harper, who in 2007 trumpeted his intention to make Canada a 'global energy powerhouse.' Trudeau's government has advertised its climate policy as one of 'clean growth' – replacing dirty coal with natural gas for both domestic and foreign consumption, implementing a modest tax on the consumption of carbon energy, pledging to eliminate 'inefficient' fossil fuel subsidies and taking steps to incentivize electric cars and other carbon-reducing technologies. These initiatives 'put Canada on the map as one of the few countries in the world (outside of Scandinavia) with serious climate legislation based on pollution pricing' (Lourie 2024). Yet no initiatives, whether at the federal or provincial level have directly addressed the elephant in the room: the outsized fossil industry, which in Canada is centred upon the extraction of bitumen from the tar sands of Alberta and Saskatchewan. Instead, Canadian governments have been busy approving, subsidizing and even purchasing massive pipeline projects to move bitumen (and LNG) to global markets. As part of ongoing settler colonialism, the Canadian state has repressed protestors who have attempted to stop these projects as they intrude onto unceded Indigenous land. According to official statistics, Canada's greenhouse gas emissions stood at 761 Mt Co_2 in 2005, which is considered the 'base year' for its emission reduction target. By 2019, the country's emissions were measured at 752, a reduction of 0.012 percent in 14 years. Overall measured emissions fell to 686 in 2020, due to the COVID-19 recession (Canada 2024). But by 2022 emissions had risen to 708, as the treadmill of production regained speed. In the same year, Canada's federal government announced a plan 'to slash emissions by 40 to 45 percent below 2005 levels by 2030' (Tasker 2022) without addressing the elephant in the room. Trudeau did address that elephant, or rather the leading lights of fossil capital, in a speech he gave in a room in Texas in 2017, when he reassured the assembled capitalists that 'no country would find 173 billion barrels of oil in the ground and just leave them there.' (as cited in Carroll 2021:7). A year later, his government agreed to purchase the Trans Mountain Pipeline, designed to triple the capacity to transport bitumen to the west coast, but in financial terms, an elephant of a different colour, namely white (Bakx 2024).

Export-oriented fossil fuel production is strategically important in Canada's 'Clean Growth and Climate Change' plan. David Hughes (2024) points out that 'in 2022, these fuels met 77.4% of the nation's end-use energy demand, and the country exported 63% of its oil, 34% of its gas, and 67% of its coal production' (Hughes, 2024). Those exports are crucial to the accumulation of fossil capital in Canada, yet, importantly, they are not attributed to Canada in standard carbon budgeting, but to the countries where they are burned. By means of this accounting trick, capitalism in Canada can continue to extract and export carbon, in increasing quantities, expanding its actual carbon footprint while appearing to be on a path towards 'net zero' emissions, which helps explain why Canadian negotiators have been particularly keen advocates of 'net zero' at COP meetings (Vitello 2019). The chicanery of official calculations of this sort, and of 'net zero' as a fetishized concept (to be discussed later in this chapter), are all part of hegemonic impression management, contributing to the odd figure of a climate leader that is actually a laggard.

In fact, Canada has ranked near the top of all countries in its per-capita carbon emissions,[1] in state subsidies to fossil capital (Levin 2024) and in financial support by Canadian banks for carbon extraction (Oil Change International 2023). As the organic crisis of Fossil Capitalism deepens, a new climate regime has been emerging: capital and its organic intellectuals subtend a *regime of obstruction* (Carroll 2021), strategically shifting towards reformist measures – variously branded as 'clean growth', 'green growth' and 'climate capitalism' – that protect capital's 'economic nucleus' while appearing to address a visibly worsening climate crisis. At the Corporate Mapping Project, we mapped the architecture of the regime, focussing on both the position of fossil capital within the wider corporate power bloc and on how the power of fossil capital reaches into the state and civil society.

Regarding the former, a network analysis of interlocking corporate directorates revealed how a tightly integrated, Calgary-based fossil-capital elite is nested within the national corporate elite, with further elite ties to the transnational corporate network. The directorates of companies producing and distributing fossil fuels interlock extensively with each other and with financial and other corporations based in Toronto and to a lesser extent Montreal and Vancouver. I concluded that the architecture of corporate power into which fossil-capital firms are closely stitched, resembles 'an entrenched oligarchy' (Carroll 2017:254). The elite network complements a network of capital relations in which wealthy individuals, financial institutions (including asset managers) and corporations own significant blocs of corporate shares Carroll et al. 2021 (Carroll and Huijzer 2021), along with extensive credit relations through which financial institutions enable fossil-fuel investment (Hudson and Bowness 2021).

Regarding the latter, we found that a panoply of relations and practices protects 'revenue streams issuing from carbon extraction, processing and transport while bolstering popular support for an accumulation strategy in which fossil capital figures as a leading fraction' (Carroll 2022:198). These include (1) the networks of interlocking governance boards that link corporate leaders with institutions and organizations in civil society and the state (Carroll et al. 2021; Carroll et al. 2020); (2) the network of corporate lobbyists, through which a small army of professionals communicate regularly with elected and appointed political leadership (Graham, Carroll and Chen 2020); and (3) networks of online 'extractive populist' communication that builds support for fossil capital (and hostility towards climate activism) at the grassroots (Neubauer et al. 2023).

To be sure, in Canada and elsewhere, the regime of obstruction operates at interlinked scales: in everyday life (e.g., Huber 2013), in local communities (Eaton and Enoch 2021), in institutions such as universities (Adkin et al. 2022) and in sub-national (Steward 2017), national (Graham et al. 2020) and international (Sapinski 2016) governance. The regime reproduces Fossil Capitalism economically while intervening in political and civil society to align popular and political sentiments and understandings with the interests of fossil capital (see Table 4.1).

Each national economy holds a specific, historically evolved position within global capitalism. In Canada, the abundance of low-grade bitumen has led fossil capitalists, supported by the state, to erect massive extractive facilities primarily in Alberta, with equally massive networks of oil pipelines – fixed-capital investments that must be utilized until they are depreciated, if they are to release their value as profit. In the current conjuncture, the shift towards renewables, although 'too little too late' as a climate-change mitigation strategy (more on this below), presents a looming threat of 'stranded assets.' If demand for petroleum flatlines or even falls, the enormous

TABLE 4.1 Multi-scalar hegemony of fossil capital

Scale	*Key instances*
Everyday life	Fossil-fuelled consumer capitalism as a way of life: the privatized geography of automobility and suburbanization
Local community	Civic privatism/boosterism and hegemonic community economic identity: 'The oil industry is us', etc.
Institutions	Entrenchment of fossil interests in institutions of knowledge production, etc.: petro-universities and state-subsidized R&D
Sub-national	Fossil boosterism in extractive and sacrifice zones
National	Contention around the 'national interest', through elite policy-planning and online extractive-populist networks
Transnational	Global elites and policy-planning/governance

Adapted from (Carroll 2021:482).

fixed-capital investments will turn into bad debts and likely bankruptcies – an eventuality particularly applicable to Canada's tar sands with their high production costs and low-quality product (Belot 2024).

> The climate crisis makes it increasingly likely that a significant portion of global hydrocarbon reserves must be left in the ground and never valorised as commodities. There is thus a geo-economics of oil, in which producing countries push to maximise extraction and ensure the realisation of the value of their resources. While limiting production increases prices, it also increases the likelihood that a particular reserve will remain buried.
>
> *(McCreary 2021:50)*

For political and economic leadership in Canada, the calculus is that the '173 billion barrels of oil in the ground' about which Trudeau boasted in 2017 must be brought to market as quickly as possible, before its value disappears. Thus the frenzy of pipeline approvals. But coupled with that is an emergent emphasis on developing and deploying the technology of 'carbon capture, usage and storage' (CCUS), which promises to reduce carbon emissions at source. Already heavily subsidized by the Canadian state (Woodside 2024), CCUS is one of several types of geoengineering technologies catering to the dream of an incremental transition from carbon-based to post-carbon energy systems.

The regime of obstruction we studied in Canada has been broadly matched by developments elsewhere, such as Australia (Wright and Nyberg 2015). In the United States, the 'climate denial machine' engages in 'academic capture' through increasing financialization of higher education, funding and influence from the fossil fuel industry, and the reticence of university employees to challenge the status quo (Lachapelle et al. 2024). Alan Betts describes the regime of obstruction there as the 'Fossil Empire', noting that the obstruction goes back to the 1980s, when Exxon repressed its own research findings (which correctly forecast the coming climate crisis) and, by 1989, led in creating the Global Climate Coalition which disputed the scientific basis for unease about climate change. Exxon was also active in preventing the United States from endorsing the 1998 Kyoto Protocol, the first international treaty on climate.

> One of the central reasons why there has been so little progress in moving away from burning fossil fuels, despite IPCC Assessment Reports going back to 1990, has been the continued opposition of the fossil fuel companies, most of it hidden from the public. Strategies for this deception have changed with time. In recent decades a conservative climate change counter-movement has developed both in the United States and a similar one in Europe which merges the interests of the Fossil Empire and neoliberalism in delaying government and international action on climate change. One strategy uses media articles and advertising to suggest that individual and corporate actions are sufficient. ... In the past decade as climate disasters of many different kinds have mounted around the world, simple denial looks obviously dishonest, so the Fossil Empire strategy has shifted to deflection. This strategy has been widely used to shift issues from corporate responsibility to one of individual personal responsibility. ... Climate activists can be encouraged to argue over whether one should travel less, buy an electric car, install solar panels, eat a vegetarian diet, have fewer children or live a simpler life. The list is endless. This strategy has been very successful in deflecting attention from what needs to be done at a collective level to the individual level, where people can either feel they are taking useful steps, or instead perhaps feel guilty.
>
> *(Betts 2021: 212; cf. Brulle 2018; Dunlop and McCright 2015)*

Moreover, just as in Canada, where we found financial capital to be closely aligned with fossil capital, Adrienne Buller (2022:102) observes in a more global context that 'a relatively small cohort of firms has come to form a new locus of economic power by two connected means: by amassing and concentrating ownership across the economy, and in the process of doing so fundamentally altering how capital is allocated and the future constructed.' She calls our attention to the giant asset managers like BlackRock and Vanguard whose ownership and credit relations with most of the world's large corporations 'are fundamentally altering how finance understands, responds to, and ultimately constructs the future' (ibid.). This shift in financial architecture, which has gained pace since the 2008 meltdown, constitutes a quantum leap in the process of global financialization, which in turn has conditioned the 'solutions' to the climate crisis proffered by hegemonic institutions.

From First-stage to Second-stage Climate Denialism

Across the global North, regimes of obstruction evidence the same basic pattern of concentrated corporate power blocking the changes that might be effective given the scale and scope of the climate emergency. A common ideological element in all cases is climate change denial. However, as the crisis has deepened, regimes of obstruction have evolved from outright denial of the causal relation between burning carbon and climate change to a hegemonic strategy I identified earlier as green growth or clean growth – phrases that wed ecological wellness with capital accumulation.

As Charles Derber has suggested, the 'denial regime' (an assemblage of fossil capital, its political allies and the 'think tanks, policy institutes, and media outlets that offer a scientific patina of respectability for denial' (2010:75)) has shifted from 'stage 1' to 'stage 2.' Whereas in the first stage global warming was dismissed as a hoax – the dismissal giving license to 'Drill, Baby, Drill!!!' (Derber 2010:76) – stage 2 denial, coming onto the scene in the mid-to-late 2000s, is more insidious. The human causes of climate change are accepted, but the response must be very gradual so as not to disturb existing investments. 'The secret to stage 2 is to propose lines of action that appear to be credible responses to the truth now officially acknowledged but do not run the risk of hurting big oil and coal companies or toppling the entire capitalist applecart' (Derber 2010:82).

At the Corporate Mapping Project, we followed Derber but recast the denial regimes as 'traditional' and 'new', pointing out that the latter has not disappeared. Rather, the two forms of denial reinforce each other in a complex but versatile hegemonic project that structures climate politics 'around an apparent divide between the reactionary conservative–populist forces of outright denial on one side, and a more progressive-leaning incremental agenda for action on the other' (Carroll et al. 2022:223–4). The project is in a sense self-contradictory, yet well-suited to postmodern forms of hegemony as 'consent without consensus', which emerged during the decades of deepening climate crisis (see Chapter 3). Diana Stuart and her colleagues have described the basic modus operandi of new denialism in the United States:

> Fossil fuel companies, especially those identified as the 'carbon majors,' are spending an increasing amount of money to block climate policy while misleading the public. A recent report found that fossil fuel companies spend around $200 million each year to block meaningful climate policy through lobbying and an additional $195 million each year on advertising campaigns that falsely suggest they are devoting significant funds to green initiatives (Laville 2019). Others have found that fossil fuel companies will publicly support alternative energy and efforts to reduce GHG emissions while simultaneously lobbying to undermine climate legislation.
>
> *(Stuart et al. 2020:439)*

Whether they work within the corporate, state or civil-society sector, proponents of new denialism diverge from traditional denial in accepting the science of climate change and advocating an active policy response. But the response consists of 'a limited agenda for action that does not threaten capital accumulation by the fossil fuel industry' (Carroll et al. 2022:222). In the process, the meaning of 'climate action' becomes ambiguous, allowing 'industry and governments to create the illusion of action – whether through the adoption of voluntary emissions reduction measures or incremental policy action' (ibid.).

Climate Capitalism, Green Capitalism, Clean Growth...

In fact, stage 2 or new denial is shorthand for a raft of 'false solutions' to the climate crisis, now favoured by regimes of obstruction. The solutions are informed by ecological modernization theory, a sociological framework introduced in the 1990s which views 'the incorporation of nature' into 'the capitalist economic process' as increasingly feasible and 'in some respects even desirable' (Mol 1995:42). The basic premise is that 'the climate crisis can be transcended through transitioning out of a fossil fuel-based energy regime and into a post-carbon energy regime within the confines and logics of continued capital accumulation' (Chambers 2021:115).

Climate Capitalism is, of course, a class project. Kevin Surprise has identified a growing awareness, within the ruling class, of the threat that climate breakdown poses for its own dominance.

> Capital is beginning to recognize its ecocidal propensities as fundamental threats to continuing hegemony. Of the top 10 'most severe risks' in the World Economic Forum's (WEF) 2022 Global Risks Report, five are labeled 'environmental,' with the top three being climate change, extreme weather, and biodiversity loss. Indeed, the climate crisis has been identified as a fundamental threat to the financial stability of capitalism by a range of elite institutions.... As capital destroys the planet, leading capitalist actors and institutions are beginning to recognize that this may augur severe threats to the maintenance, reproduction, and dominance of capitalism as well. The primary response to these contradictions has been so-called 'green capitalism': attempts to transition to sustainability via renewable energy markets and investment, carbon markets and offsets, privatized conservation, natural capital, green bonds, green consumerism, bioenergy, carbon capture and storage, carbon removal, and so on without fundamentally altering capitalist forms of class domination, exploitation, and accumulation.
>
> *(Surprise 2024:455)*

Following John Foster (2010), Gunderson, Stuart and Petersen argue that this paradigm has at its core two forms of fetishism which are themselves 'strategies for denial':

> Technological fetishism is seen in the prevalence of strategies to increase the efficiency of technologies as well as an extreme version in recent calls for geoengineering. Market fetishism is seen in the rising popularity of market-based climate change mitigation policies, namely cap-and-trade schemes.
>
> *(2018:134)*

Articles 6 and 10 of the 2015 Paris Climate Agreement that resulted from the 21st Conference of the Parties of the United Nations Framework Convention on Climate Change (COP 2015) exemplify these strategies. Article 6 introduced the notion of an international carbon market and established what has become known as the 'Sustainable Development Mechanism', which allows industries in the global North to invest in emission reduction efforts (carbon offsets) in the global South, in exchange for permission to increase carbon emission caps. Article 10 of the Agreement establishes a 'Technology Mechanism' to increase climate technological innovation in reducing GHG emissions and promoting economic growth (Gunderson et al. 2018:135).

Technological and market fetishism frame the concept of 'Climate Capitalism', also known as 'green capitalism', 'green growth' and 'clean growth.' Whatever the moniker, these approaches 'all propagate a mix of technocratic-technological-financial-market solutions for all central issues concerning the climate crisis, neglecting to address 'the existing unsustainable politico-economic arrangements,' reviewed in Chapters 1–3, which are driving the crisis (Ilc, 2021:345). Committed to 'harness[ing] the forces of capitalism to tackle the climate problem' (Rathi 2024:2), Climate Capitalism is grounded in the assumption that to deal properly with the climate crisis, high growth rates are needed. As growth creates new centres of accumulation around renewable energy and the like, the world can transition to sustainability without disturbing capitalist accumulation (Surprise 2018:1233). Paradoxically, the project is based on the assumption that 'the main cause of the crisis, economic growth, is crucial to the solution of the very same crisis' (Buch-Hansen and Nielsen 2023:357). The tight link between capital accumulation and climate breakdown suggests that it is not possible 'to break the connection between economic growth and emissions on a global scale and in the time available' (ibid.). This is why Climate Capitalism fits within the scope of stage 2 denial, offering false solutions to the climate emergency.

In Adrienne Buller's view,

> Green capitalism is an effort to address environmental catastrophe through new paths to accumulation while minimising disruption to our current economic systems and modes of living, irrespective of whether the actions taken actually lessen the damage they claim to or cause other harms in the process. Green capitalist solutions are predicated on the continuation of the destructive processes, systems and economic relations that have both delivered us into this state of crisis and severely delayed action to stymie it. *(2022:viii)*

She goes on to describe green capitalism as the union of two 'defining pillars.' First, an effort to preserve capitalism while addressing its tendency (discussed in Chapter 1) to destroy its own conditions of production. Second, an effort to foster new fields of accumulation in transitioning to a decarbonized, ecologically sustainable economy. Green capitalism is, in this way, 'a response to the collision of the accumulative drive of capitalist economies with the profound threat to returns, asset values, and accumulation posed by deepening ecological crisis' (2022:232), which endeavours to transform the threat of a dying planet into new opportunities for profit. Advocated by a raft of hegemonic institutions, from the United Nations and its COP initiative to capitalist-run outfits like the World Economic Forum, Climate Capitalism is 'presented as both a necessary and a pragmatic strategy for "saving" nature based on the rationality of prices, markets and capital' (ibid.). Saving nature while saving capitalism from itself. What's not to like?

Climate Capitalism follows in the grooves of neoliberal hegemony, privileging markets and the creation of profit opportunities over state-led planning. Within this policy mix, however, the state still plays an active role 'as a facilitator of new market domains and as a "de-risker" of private capital', as public policy both safeguards and shepherds capital 'into previously undesirable areas through a heady blend of market making, incentive, and guarantee' (Buller 2022:274). Green capitalist solutions, then, add up to a privatized response to the climate crisis, seeking 'to transfer the complex, ethically and socially fraught, and inherently political questions presented by ecological crisis from democratically contestable terrain to the private authority of markets, with outcomes ultimately driven by the self-interest of rational actors motivated by profit' (ibid.).

Within the bounds of capitalism, the solution to climate crisis lies in bringing decarbonizing technologies directly into commodity production and consumption, using market mechanisms to incentivize climate-friendly investment. By decoupling accumulation from carbon extraction – in particular, by changing the energy source – economic growth metamorphoses from problem to solution. It becomes 'the very foundation of the innovation and entrepreneurship required to fix things' (Daggett, 2018:33–4). All the while, market-based solutions steer the rational choices of investors and consumers towards ecological modernization, without requiring significant changes to lifestyles. Buller (2022), however, notes one major deal-breaker: 'Ecological crisis is not a problem for which we have the luxury of time for price signals to gradually shift economic activity.'

In the remainder of this chapter I take up Climate Capitalism's twin pillars, first ecological modernization as a techno-fix, then carbon pricing/taxing/offsets as a market-based fix. The chapter's final section returns to ecological modernization to consider the ambitious notion of geoengineering through carbon capture, utilization and storage (CCUS) and solar geoengineering.

Ecological Modernization

As the climate crisis has worsened, the most obvious instances of ecological modernization in everyday life have appeared in retail commodities like light bulbs in the 1990s and, more recently, electric cars. If, as manufacturers produce greener goods, consumers in their minions can be led, through generous state subsidies or through taxes on carbon-heavy purchases, to make the smart choices – for energy-efficient light bulbs, heat pumps and hybrid cars or, better still, fully electric cars powered by rooftop solar panels – a massive reduction in carbon emissions will ensue. To be sure, creating a liveable world from the wreckage of consumer capitalism will require shifts away from high-emission products and practices. But concerted efforts through market incentives and promotional advertising to replace the internal-combustion engine (ICE) car, Fordism's poster boy, with the electric car illustrate how ecological modernization is ensconced within the trifecta of accumulation, imperialism and hegemony. The implication is that an 'ecological switch' (Castree and Christophers 2015:380) that simply changes the energy source from carbon to electrons will not solve the earth crisis.

Vaclav Smil (2022) reminds us that electric cars are a stunning example of the 'material dependencies' in energy conversions, requiring 'larger masses of old materials as well as unprecedented quantities of materials that were previously needed in only modest amounts.' A typical lithium car battery weighing approximately 450 kilograms contains lithium, cobalt, nickel, copper, graphite and other materials which are refined from 'extracting and processing about 225 *tons* of raw materials' (Smil 2022, emphasis added). Smil cites estimates that, were electric cars to comprise 50 percent of fleet vehicles by 2050, lithium extraction alone would have to grow by a factor of 20 (i.e., 2000 percent). All these extractive and refining activities have their own carbon footprints, making electric cars much 'dirtier to build' than ICE cars. The climate benefits of electric cars only begin after thousands of kilometres have been clocked (DeSmith 2023), although details on the size of an electric car's 'carbon debt' are disputed (Evans 2023). Looking further into the electric-car commodity chain, we can see that such automobiles are operationally emission-free only if the electricity they consume is not fossil-powered. Yet worldwide, the carbon intensity of electricity in 2022 stood at 438 grams of CO2-equivalents per kilowatt-hour of electricity generated, with coal-powered countries such as South Africa requiring 708 gCO_2 while hydro-/nuclear powered countries like Canada require 126 gCO_2. For most drivers, the electric car generates carbon emissions, and although renewables are gradually replacing fossil fuels in many electric grids, as of 2021, when 50 percent of the increased global demand was met by fossil fuels, renewables contributed less than a third of global electricity generation (Energy Institute 2024). Moreover, renewables themselves have massive footprints on the production side, and are more appropriately termed 'rebuildables' since to 'recycle' them requires considerable labour and (typically fossil-sourced) energy (which is why solar panels and wind turbines are conventionally dumped into landfill once they have depreciated; Friedemann 2021:70). Electric cars are not as green as they may appear to be. Just as importantly, as a 'solution' to climate crisis they tend to promote a transportation system based on solo cars, requiring 'a complex infrastructure of bridges, roads, highways, etc. that's highly polluting (the production of concrete and asphalt is a major industrial source of GHG emissions), all at the expense of less damaging alternatives like public transit and cycling' (Legault et al. 2023:85).

In their discussion of the recent 'ecological modernization of automobility', Brand and Wissen (2021:154) aver that

> the 'ecologization' of automobility through the market-based and technologically fixed strategies ... is an attempt to perpetuate the imperial mode of living through a selective ecological modernization of one of its central domains. The crucial questions are rarely asked in the prevailing debate over a 'mobility transition': how could transport be avoided and traffic routes shortened, and how could the necessary transport be organized in order to be socially and environmentally sound?
>
> *(2021:158–9)*

As the example of the electric car shows, ecological modernization operates within the logic of consumer capitalism to *mobilize the consumer*. Green consumerism is aimed at middle-to-upper class people 'who are eager to reduce their CO_2 footprint by participating in atomized acts of consumption that do not interfere with the system responsible for those emissions' (Dockstader and Bell 2020:657). However, ecological modernization's claims are contradicted by empirical scholarship that 'demonstrates that green consumerism and capitalist growth cannot reduce emissions on the scale required to avert climate catastrophe' (Dockstader and Bell 2020:655). Stuart et al. (2020:443) add that individual-level approaches to 'escape from carbon-intensive lifestyles, even if adopted widely, will not adequately address climate change.' As a reminder that Fossil Capitalism is the project of the capitalist class, they refer us to Richard Heede's carbon accounting, which has shown that, since 1988, '70% of carbon emissions can be traced back to just 100 companies' (ibid.; Griffin 2017). The refusal of ecological modernization, again, to address the elephant in the room is plain. Indeed, in a world pierced by ever-deepening inequities, a recent book by billionaire Bill Gates, a champion of ecological modernization, acknowledging that technological solutions are not entirely adequate, mentions the word 'innovation' 90 times, without a single mention of 'inequality' (Buller 2022:191).

The electric car provides a good entry point into a wider discussion of ecological modernization. We can see how the car itself, and all the ancillary industries it boosts along its commodity chain (lithium extraction being only one), offer tremendous opportunities for accumulation. Moreover, its promotion and marketing contribute to a neoliberal hegemonic project that minimizes the need for structural change in addressing the climate crisis, placing the burden mainly on the ecologically conscious consumer and the governments that strive to incentivize the 'smart choices' by investors and consumers alike. In the shadows of this scenario, to be illuminated presently, is the third piece of our trifecta of power: imperialism.

Before we get to that, I must take up what is really the master thesis of ecological modernization as it applies to the climate crisis: the notion of 'decoupling' energy. Put simply, decoupling means that through technological innovation an economy can produce more commodities using fewer raw materials (including carbon energy) and creating less pollution. In effect, decoupling would reduce both withdrawals from ecosystems and externalized deposits, replacing the treadmill of production with a 'circular economy' reliant less on primary resources and producing less waste. As it pertains to Fossil Capitalism, green growth means decoupling the economy from fossil fuels, and thereby 'decarbonising our profoundly unequal present while preserving, to the best extent possible, the governing logics, structures and infrastructures that sustain it' (Buller 2022:239). In itself, decoupling is a worthwhile, indeed crucial objective. But is it a feasible objective within capitalism, on a planet with finite resources? As we saw in Chapter 1, the profit motive at the centre of capitalism compels capitalists to reinvest their profits in competition with other capitalists, creating an intransigent dynamic of endless growth. Can such growth be green, clean? This is what decoupling promises.

In the 19th century, economist William Stanley Jevons showed that efficiency improvements in the use of coal made coal more cost effective and therefore more attractive to consumers, who chose to burn more coal.

He concluded that 'greater efficiency in resource use often leads to increased consumption of resources' (Clark and York 2005:411). This relationship, known as the Jevons Paradox, has bedeviled ecological modernization. Instead of decoupling the economy from the ecosystem, 'technological "improvements" have actually increased the amount of resources used, since expansion in production typically outstrips gains in efficiency' (Clark and York 2005:391). An added challenge is that non-renewable resources are, by definition, subject to depletion over time. Extraction moves 'from the most accessible to the most out-of-reach resources, and more land area per unit of resources is extracted' (Legault et al. 2023:84). Oil, of course, is a key example, so it is not surprising that 'the ecological conditions that allow oil to be extracted have greatly deteriorated in recent decades. It takes much more energy to extract a barrel of oil today than in the past' (ibid.). Given that most of the world's highest quality oil has already been extracted, corporations have turned to 'extreme oil' alternatives such as bitumen, which requires enormous (mostly fossil) energy input and yields lower energy returns. An instructive case is that of Alberta's tar sands. While the Energy Return On Investment (EROI) is 25:1 for conventional oil (25 units of energy produced for every 1 unit of energy input), bitumen extraction shows a ratio between 2.9:1 and 5:1 (Nuwer 2013). The overall picture is a 'plummeting energy return on investment of oil', raising the prospect that 'in the foreseeable future, the energy needed to produce oil liquids could approach unsustainable levels, a phenomenon called "energy cannibalism"' (Misra 2023).

A few global North countries have been touted as climate leaders in decoupling GDP growth from growth in fossil-fuel consumption (Ritchie 2021), although much of the progress reflects an 'offshoring' of carbon emissions, as high-emitting industries move overseas. Globally, 'between 2000 and 2014, both carbon dioxide emissions and global GDP grew at an average rate of 2.8% per year, in almost perfect lockstep' (Buller 2022:240; cf. Pineault 2023). More recent data show a regionally uneven pattern, with the global North showing signs of decoupling (partly reflecting offshoring of production) while in southeast Asia and the Middle East – core areas, respectively, for industrial production and carbon extraction – the monotonic relationship between growth in GDP and growth in emissions continues unabated (Singh 2024). Of course, *slowing* the *additional* emissions of carbon into the atmosphere is not a viable strategy for avoiding climate catastrophe. What is needed, for successful ecological modernization, is to break the relationship between economic growth and carbon emissions (ibid.). The most recent data show that the relationship persists, as 'global energy-related CO_2 emissions grew by 0.9% or 321 Mt in 2022, reaching a new high of over 36.8 Gt' (International Energy Agency 2023), making a mockery of the goal of achieving 'new zero' emissions by 2050.

Market-based Solutions?

Within the neoliberal policy paradigm that has been hegemonic since the 1980s, the pathway to ecological modernization must hue to a market-based course. Key to this is the concept of 'natural capital', which assigns a monetary value to each input to production – whether soil, forest, water, mineral resources or species. This internalizes 'nature' within the accumulation process, as financial markets 'digest' 'the complex entanglements of social and "natural" relations' that actually comprise our world, reducing them to a price (Hudson 2021:107). Fressoz and Bonneuil (2017:65) reveal the logic at work in this practice:

> The 'invisibilization' of the limits of Earth is no longer just a result of an externalization (as a great outside that humans can draw from and jettison into without problem), but on the contrary of a radical internalization of earthly entities and processes into the realm of financial markets. This internalization is expressed in the efforts to measure ecosystemic functions in terms of financial flows, conceptually creating a nature that is liquid and capitalizable even in its most intimate processes.

If nature can be internalized in this way, 'the existing economy can remain as it is' (Brand and Wissen 2021:164). Yet in reducing complex ecosystems to the fetishized category of natural capital what are lost are the 'many interdependencies in society and nature' that '*cannot be expressed in terms of prices*' (Altvater 2016:151). These interdependencies include the consequences of climate change – human suffering, extinction, erasure of some landscapes in favour of others – all of which vanish 'under the streams of numbers that are the only actionable information markets can handle, since they are among the abstractions upon which exchange rests' (Hudson 2021:107).

This 'internalization' is often presented as 'putting a price on carbon', thus integrating 'natural capital' into the calculations of market participants. Once appropriately priced, market actors incentivized by profit, will shift from high-emitting activities. A steep price on carbon, reflecting its real impact, 'will necessarily push profit-seeking firms and capitalists to innovate and adapt, and individuals to shift to decarbonised sources of energy' (Buller 2022:30). Yet although carbon markets have been celebrated in the European Union since 2005, Adrienne Buller goes on to ask 'why has no jurisdiction managed to enact a carbon price effective enough to bring their economy in line with the trajectory of a safe future?' (2022:31).

Clearly, carbon markets, which can take two forms – carbon taxes as in Canada and cap-and-trade systems as in the EU (Buller 2022:59) – require extensive state participation, yet that participation is downplayed within the governing ideology of Climate Capitalism. The state's creation and management of 'natural capital' actually reinforces neoliberal hegemony, depoliticizing the socio-ecological issues raised by decarbonizing initiatives.

> By establishing rights to pollute the atmosphere that can be bought, sold and converted into novel financial assets, new pathways for profit and speculation are created, while the need for states to impose meaningful regulation can be met in a cosmetic sense without the risk of alienating the corporate and capitalist base to which they are in various ways beholden. In the process, all the political and value-laden questions associated with decarbonisation – improving public health, escaping our dependence on fossil fuels, limiting further environmental impacts – are reduced to the singular goal of curbing emissions in order to make the question of governing emissions into a market-compliant unit.
>
> *(Buller 2022:76)*

The buying and selling of the right to pollute often takes the form of 'offsets', which can be purchased to provide monetary compensation for environmental damage that issues from production (e.g., carbon extraction) and consumption (e.g., air travel). There is a strong hegemonic aspect to this accounting practice. 'Designed as an incentive to reduce environmental pollution, offset markets institutionalize the influence of forces whose economic success depends on the continuation of the imperial mode of living, albeit in an ecologically modernized form' (Brand and Wissen 2021:172). The idea is that

> if private individuals or companies can pay 'indulgences' in the form of a variety of offsets for the environmental damage of consumption and production, then a broadly shared conviction may emerge that nature is in principle replaceable – emissions in one place are offset by reforestation measures elsewhere; ecosystems that are annihilated for a motorway junction in one place will be restored in another; a vacation flight ceases to be ecologically suspicious if you pay for a tree to be planted that will absorb the emissions generated by your flight over the course of its life cycle. The idea of neoclassical environmental economics, according to which 'natural capital' can be easily replaced as long as the total capital stock continues to grow, thus becomes common sense.
>
> *(Brand and Wissen 2021:173)*

This reasoning, which underlies all offset schemes, flies in the face of basic ecological science, which recognizes the dynamic, biologically diverse interdependencies that characterize healthy ecosystems. Breaking these systems into discrete, costed units sets us on 'a path toward a possible future of pristine mono-species forests – a sanitised idyll worthy of a theme park, and devoid of the ability to sustain diverse life' (Buller 2022:251).

Climate Capitalism and Eco-imperialism

What, then, of Climate Capitalism's implications for the third piece of our analytic trifecta? As Jayati Gosh and her colleagues have observed, 'the development of new technologies has never provided a route out of imperialism ... but it can and does change the nature of the resources that are sought to be controlled by the major powers.' (Ghosh et al. 2022: 80). To be sure, there is an unmistakeable geography to the creation of 'natural capital' within market-centred ecological modernization schemes. On an industrial scale, 'green energy' is often devastating; it is 'completely tied into some of the biggest colonial land grabs in the twenty-first century: Lithium mining, wind parks, solar farms' (Gelderloos and Dunlap 2023:5). Michael Albert offers some telling examples of the 'detrimental impacts of renewable energy supply chains' on communities in the global South when social justice concerns are ignored: 'in the horrific cobalt mining conditions in the Democratic Republic of the Congo; accumulating toxic waste from solar panels, electric vehicle parts, and smart meters in Ghana; and processes of land enclosure and dispossession of marginalized communities to build solar energy plants in Gujarat, India' (2021:93). He concludes that, within a regime of endless growth, shifting to such post-carbon energy 'would require dramatically increasing land use and rates of mineral extractivism, which would intensify processes of land enclosure, dispossession of marginalized communities, contamination of water resources, RE-technology waste proliferation, and biodiversity decline' (2021:94). The conversion of 'nature' into 'capital' might not serve the needs of humans and the ecosystems in which they are embedded, but it does serve a crucial purpose within a capitalism struggling to manage a dual crisis, both ecological and economic. At its heart, Buller observes, Climate Capitalism 'opens up vast new terrains for the expansion of capital into the natural world. And for no industry has this proven more appealing than finance, which has seized upon the speculative prospects of this new frontier' (2022:255).

These vast new terrains are typically in the global South, supporting Brand and Wissen's thesis that the imperial mode of living requires 'elsewheres' that form the invisible premise for an unsustainable way of life in capitalism's core. In fact, carbon markets and offsets are integral to 'new corporate enclosures.' To attract investors, carbon offsets must be accompanied by socio-ecological relations that 'ensure a pristine carbon sequestering landscape'; hence 'carbon market governing regimes require landscapes and "nature" (including trees and carbon) to be "hemmed in" so as to minimise leakage and ensure permanence' (Richards and Lyons 2016:211). As with earlier enclosures in the history of capitalism, the new corporate enclosures entail eviction and dispossession, often legitimated as 'opening up' land to 'green development' as a positive outcome for local communities (ibid.). Belying its benign appearance as a market-fix, the financialization of nature amounts to 'a global movement of appropriation of land for the purpose of activities rewarded by the sale of "carbon credits," "biodiversity credits," etc., further dispossessing rural and indigenous populations from their commons' (Fressoz and Bonneuil 2017:63). Indeed, Adam Bumpus and Diana Liverman (2008) characterize the global carbon offset regime as a project of 'accumulation by decarbonisation': a telling instance of accumulation by dispossession, combining 'direct violence; land grabs and the erasure of livelihoods; new forms of commodification and financial speculation; and the enclosure of a wildly disproportionate share of the atmospheric carbon sink and terrestrial resources by powerful firms and the globally affluent' (Buller 2022:88).

'On paper', carbon offsets work because atmospheric carbon is global; hence offsets can be located anywhere. Given the reality of imperialism, the lion's share of offsets are bought by individuals, corporations and states in the global North 'in order to facilitate their continued enclosure of the atmospheric commons, while the offsets themselves involve the sequestration of land in the Global South in service of this atmospheric enclosure' (Buller 2022:77). Buller concludes that these schemes are nothing more than 'neocolonialism in its most distilled form: the forcible transfer of sovereignty from the people who occupy the land to those with sufficient monetary and coercive power to ensure it is used in their interest' (2022:101).

The topsy-turvy world of carbon offsets brings us to a key concept in Climate Capitalism, namely, 'net zero.' According to the World Economic Forum,

> the term net zero applies to a situation where global greenhouse gas emissions from human activity are in balance with emissions reductions. At net zero, carbon dioxide emissions are still generated, but an equal amount of carbon dioxide is removed from the atmosphere as is released into it, resulting in zero increase in net emissions.
>
> *(Wood 2021)*

Reaching net zero by 2050, the goal adopted at the 2015 COP meeting in Paris will require huge investments to 'scale up' new emissions-busting technologies that remain at early stages of development while bringing costs down (ibid.). Using carbon offsets as the financial catalyst, initiatives to reach net zero create 'negative emissions' to offset continuing emissions by corporations and states purchasing offsets – either by 'natural' means or through technologies (Cran-McGreehin 2021). The former involves preserving or creating natural carbon sinks, such as forests, peat bogs and coastal ecosystems. The latter uses technologies to remove carbon from the atmosphere, none of which has been proven to work at the required scale. In either case, these efforts fall under the rubric of *geoengineering*, defined by Harvard's Solar

Geoengineering Research Program as a 'set of emerging technologies that could manipulate the environment and partially offset some of the impacts of climate change.'[2] Given the small, and rapidly closing, window for avoiding climate catastrophe, IPCC projections now routinely assume that there must be massive carbon drawdowns if climate breakdown is to be avoided.

Negative emissions return us to the question of ecological modernization. As with electric cars and other technological initiatives, the problem with these approaches lies with their inadequacy in dealing with the scale of the climate crisis we are facing. Preserving and restoring ecosystems as carbon sinks is obviously a good idea, and technologies to capture and store carbon may, if proven feasible and safe, have a role to play in a comprehensive energy transition. But as they are implemented within the logic of financialized capitalism, these geoengineering approaches tend to have perverse effects. As Wim Carton (2019) points out, for capital, the point in removing carbon dioxide from the atmosphere to balance continuing emissions is actually *to defer the devaluation of fixed fossil capital*, avoiding the risk of stranded assets. Offsets enable the industrial producers and consumers of fossil fuels to continue with business as usual, profitably redeeming the full value of the mines, refineries, pipelines and other investments now in place, or planned. From the standpoint of capital, this is clearly beneficial, yet 'the promise of negative emissions could act as a deterrent to more radical near-term action' (2019:764).

More recently, Public Citizen's (2024) examination of how carbon offsets actually function confirms Carton's worry.

> The challenges with carbon offsets as a climate risk mitigation tool are manifold: The most relevant is that they do not mitigate climate change – at best, their use slows the rate at which climate change intensifies and incrementally delays tipping points; at worst, they invite fraud and prevent decarbonization of our most polluting industries, guaranteeing we will never achieve net-zero, much less 'real-zero.' With the prospect of reaching peak emissions receding into the background as war and other geopolitical forces drive increased appetite for fossil fuels, carbon credits/offsets appear much more like a dangerous distraction than a solution.

Increasingly, Public Citizen observes, 'fossil fuel companies are relying on carbon offsets to justify their transition to clean energy without making material emissions reductions' (2024). Commenting on Public Citizen's analysis, Lang (2024) points to three crippling problems with carbon offsets. First, fraud is built into carbon markets since 'it is impossible to prevent manipulation of a commodity that no one can see, which is generated by a fictional story about what would have happened in the absence of carbon finance' (2024). Second, proving a carbon-sequestration project would not have gone forward without funds from sales of carbon offsets 'requires proving a counterfactual,' opening the door to widespread fraud and 'creative accounting.' Third, guaranteeing 'permanence' – that, say, trees planted in Brazil will continue to fix the same quantity of carbon released by burning coal in Germany – is a physical impossibility for all nature-based credits.

In fact, global warming is already ruining rosy conceptions of ecological 'permanence.' Each summer in Canada, for instance, wildfires are consuming enormous areas of boreal forest, which have served as carbon sinks but are now carbon mega-bombs. In the 1990s, Canada's managed forest removed on average 160 million tonnes of carbon from the atmosphere, but by early September of 2023 wildfires had released 2 billion tonnes. *The carbon released in the 2023 wildfires amounted to three times the total emissions from Canada's carbon-dependent economy* (670 million tonnes; Cecco 2023). To say the least, when a carbon offset is marketed as preserving a forest that would otherwise be cut, or creating a new forest to fix carbon, there can be no assurance of permanence.

Plan B: Geoengineering and Enhanced Eco-imperialism

We are quickly approaching tipping points that are likely to induce catastrophic climate change – the 'hothouse earth' about which climate scientists have been increasingly sounding the alarm. Ecological modernization, as we have seen, strives to digest 'nature' into capital accumulation, creating new industries that implement new technologies that incrementally replace fossil power with renewables, or that capture carbon from the atmosphere and store it in one form or another (trees, underground, under water). Offsets, carbon taxes and state subsidies incentivize these initiatives. But it is already crystal-clear that the scale of the problem is far bigger than what all these measures can accomplish. Consider that, for instance, in 2018 the Canadian government pledged to plant two billion trees over the decade, which would have fixed two million tonnes of emissions annually. It does not appear that this goal will be met, but in any case, the projected carbon-fixing from this geoengineering amounts to 1/1,000 of the carbon released in the wildfires of 2023 (which, incidentally, is not counted against Canada's commitments under the 2015 COP agreement; Cecco 2023).

Just as the depletion of oil deposits has led to 'extreme oil', the growing urgency of the climate crisis has spawned a range of geoengineering schemes to buy time for a slow energy transition compatible with capital's accumulation needs. Negative emissions achieved through geoengineering can offset the real emissions from continued extraction and use of fossil fuels.

Geoengineering the Climate, a 2009 report by the British Royal Society presented two technological approaches to achieving negative emissions which 'were reproduced countless times over the next decade' (Buck, Sapinski and Malm 2022:4). Carbon Dioxide Removal (CDR) is 'land-based.' It captures carbon from the atmosphere and fixes it, whether through planting trees, stimulating the growth of algae in the ocean (Lopez 2023) or through Direct Air Capture (DAC, discussed below). Solar Radiation Management (SRM) is 'air-based.' In the most popular SRM scenario, sulphur-based aerosols would be injected into the stratosphere, with the intent to 'modify Earth's albedo to reflect a small percentage of incoming solar radiation' (Surprise 2018:1230). By reducing the amount of sunlight that reaches the biosphere, SRM would mask the effects of human-caused climate change. As Malm suggests, these two approaches entail different dynamics, and imply the construction of new *geo*economies:

> The air-based geoeconomy would be thin and fast in that it would make modest demands on the resources of the economy below and have an all but instantaneous impact on the climate. The land-based would be thick and slow, as it would drill deep into the resource base of the economy and only gradually – over decades rather than months and years – leave its mark on temperatures.
>
> *(2022:155)*

I will discuss each of these in turn, beginning with land-based strategies. Already in the 1990s, land-based CDR was gestating as an idea. At the negotiations that led to the Kyoto accord of 1997, the United States won a key concession: 'if it managed its forests well, it would be able to store a large amount of carbon in trees and soil which should be subtracted from its obligations to limit the burning of coal, oil and gas' (Dyke et al. 2021). In the end, the United States refused to ratify the agreement, but the seed had been planted. Land-based CDR actually has its origins in the 1970s as a technology for 'enhanced oil recovery.' For decades, corporations like Occidental have injected CO_2 into nearly exhausted oil fields, to force the remaining crude up to the surface. In the United States, tax credits to develop DAC facilities now provide incentives for companies like Occidental to ramp up this process (Malm and Carton 2021:23).

As the tax credits suggest, this approach fits well within the suite of 'market-based solutions' to achieve net zero, which generally involve extensive state facilitation. However, as CDR is groomed into a profitable industry, its value in transitioning from Fossil Capitalism to a 'green economy' is likely to weaken. Malm sees the possibility of a futures market, in which fossil companies purchase offsets from CDR firms. But

> for buyers to flock to a CDR futures market, there would have to be some demand. For what? For negative emissions to *offset* the actual emissions produced by those buyers in the setup of this scheme – meaning that such

> market-driven CDR would only, in the best of cases, *compensate* for emissions that the purchasing capitalists cause elsewhere, not *draw down* carbon in any absolute sense.
>
> *(2022:154)*

Within the logic of Climate Capitalism, CDR is not a means for reducing atmospheric carbon; it is part of a 'circular carbon economy' devoted to 'carbon management as opposed to elimination' (Malm and Carton 2021:22). Anja Chalmin (2024) reports that since 2020 more DAC companies have launched than in all previous years. This includes a massive one based in Texas and owned by Occidental Petroleum, whose subsidiary sold carbon credits in 2023 to Toronto-Dominion Bank, Amazon, Houston Astros and All Nippon Airways to finance the project, supplemented by a US$ 550 million investment from asset manager BlackRock. The Stratos Project, which 'plans to construct 30 DAC plants with a planned CO_2 capture capacity of one million tonnes per plant per year', was given further funding in August 2023 by the US Department of Environment, for up to US$ 600 million (ibid.). Commenting on this mega-project which includes 'enhanced oil recovery', Malm and Carton observe that, if through DAC, more carbon finds its way underground than is extracted and emitted,

> the net result is drawdown – voilà, carbon-negative oil! But if this is a profitable practice, it will give oil companies more money with which to expand their operations – explore, drill, extract afresh and farther afield; not a winding-down of the industry, but a new lease on life. In the words of Occidental, investing in DAC ensures that 'fossil fuels have a role in the energy portfolio of the world long term.' It means greater quantities of oil reaching the surface.
>
> *(2021:23)*

Recalling the concept of EROI, it is also worth noting, with Malm and Carton, that DAC requires enormous energy inputs, which by some estimates could amount to the equivalent of more than half of all electricity produced world-wide today (Malm and Carton 2021:28).

A complex version of CDR, combining the 'nature based' solution of planting trees with DAC, which has gained favour among geoengineers recently, is Bioenergy Carbon Capture and Storage (BECCS). The idea is to grow bioenergy crops such as palm trees, harvest them and burn them as fuel, capturing the emitted carbon at power stations and storing it underground. In theory, this scheme would create negative emissions: the bioenergy crops remove carbon from the atmosphere, and the same carbon, when burned, is pumped underground, rather than returning to the atmosphere. Yet 'BECCS, just like all the previous solutions, was too good to be true' (Dyke et al. 2021).

To make a real difference in avoiding catastrophic climate change, BECCS would need to be scaled up to industrial-scale plantations of fast growing monoculture crops, which 'devastate biodiversity' (ibid.).[3] The scaling up would be no small detail. It 'would likely require a land mass equal to '*all current cropland*,' worldwide (Malm 2022:153). Given that bioenergy crops are mostly tropical plants, the likely scenario is a new rendition of eco-imperialism. Setting up massive plantations and power plants would require extensive state leadership and planning, contradicting the neoliberal market fetish that informs Climate Capitalism. And, in contradiction with the techno-fetishism that informs geoengineering in particular, there are nagging political questions:

> Who will give up their land for the plantations? Whose electricity will be fired by the biomass? If the deep grooves of the world economy are anything to go by, land will be seized in the tropical South, from which the harvested biomass will be exported to the most advanced capitalist countries for burning. It would be ecologically unequal exchange on an epochal scale—and the proficiencies of capitalism in this regard are beyond doubt—bound to further fan the flames of resentment in the Global South: first you colonize us, then you wreck the climate, and then you colonize us again to make up for that wreckage and get some good fuel in the process.
>
> *(Malm 2022:155)*

What, then, of the other geoengineering solution – solar radiation management (SRM) to 'turn down the heat on Earth by reducing the amount of incoming sunlight'? (Buck et al. 2022:4). Kevin Surprise sees this approach as pre-empting capitalism's second contradiction by creating room for more emissions in the short term, enabling fixed-capital investments in fossil-fuel extraction and refining to be fully valorized as the energy industry gradually transitions to 'green capitalism.' Air-based geoengineering 'fits logically into emergent processes of capitalist hegemony, specifically ecological crisis management via green capitalism' (2018:1230). SRM is designed to maintain existing power relations by extending the timeframe for transitioning capitalist production to renewable energy (Buck et al. 2022:14). The Intergovernmental Panel on Climate Change concluded in 2023 that 'solar radiation modification approaches, if they were to be implemented, introduce a widespread range of new risks to people and ecosystems, which are not well understood' (IPCC 2023:19).

In elite circles, growing panic over climate change has led recently to cautious endorsement of SRM by governments and international organizations, for its promise to deliver fast, massive change in global temperature without seriously disturbing Fossil Capitalism. Among these endorsements, a recent White House report claims that 'SRM offers the possibility of cooling the planet significantly on a timescale of a few years' (as cited in Fazi 2024). These various reports acknowledge the serious risks, as 'modifying sunlight could alter global weather patterns, disrupt food supplies and in fact lead to abrupt warming if the practice was widely deployed and then halted' (Fazi 2024). But the possible benefits of a quick fix, which amounts to an experiment at planetary scale in real time, are beginning to seem too attractive to resist, if one operates within the logic of Climate Capitalism.

Interestingly, SRM sits uneasily within the hegemonic policy paradigm. As I discussed in the previous chapter, despite its having been discredited by the 2008 financial crisis, neoliberalism continues to be the go-to framework for state initiatives, as shown in the continuing popularity of market-based solutions to the climate crisis. Even CDR schemes, which require massive capital investments, can be brought within carbon-credit exchanges. But SRM would operate at global scale. As Malm avers,

> *If solar geoengineering is to work, it has to be operated through central planning*. Once the assumption of a central planner – one guided by reason and good intentions to boot – is articulated, all the worms come crawling out of the can. With a free market for SRM enterprises, one fleet could undo the work of another.
>
> *(2022:149)*

The market anarchy that neoliberalism has celebrated, and which is actually central to capitalism as a market-centred society, simply cannot work when it comes to this form of geoengineering. However, the policy response to the 2008 financial meltdown offers a clue as to how, as climate breakdown worsens, the hegemonic bloc might respond. The response temporarily suspended neoliberal precepts, with massive state-led intervention, including for instance bailouts of banks 'too big to fail' and temporary nationalization of bankrupt automobile corporations in the United States. During the COVID-19 pandemic, state managers again elected to put neoliberalism into suspended animation, since 'market-based solutions' to a global public health emergency are an obvious absurdity. Neoliberalism's 'success' in recent years has been accomplished through policy gymnastics that follow the rule: 'neoliberalism forever, except when a crisis requires its temporary abandonment.' My guess is that, absent alternatives articulated in struggle from below, political and economic elites may appeal to the same 'rule' regarding SRM.

Those elites include the leading lights of American capitalism – billionaire philanthropists like Bill Gates and Jeff Bezos, who strongly support SRM. Basing his diagnosis on research by Surprise and Sapinski (2023), Fazi (2024) sees 'the emergence of a "climate power bloc" encompassing liberal-technocratic politicians, certain climate scientists, environmental NGOs, "green" philanthropists, and Silicon Valley "climate capitalists." At this intersection of ideology, class and economic interests, extreme and ambitious ideas such as solar engineering find fertile ground.'

Why False Solutions Are False: Climate Capitalism as Passive Revolution

In this chapter I have reviewed the collection of responses to climate crisis that comprise Climate Capitalism – a congeries of market-based and technological measures collected under the mantle of ecological modernization. What makes these solutions 'false' is not their ineffectiveness *per se*, but the profound limits to their effectivity within a way of life committed to endless 'growth' in the form of profit-driven capitalist investment. The climate crisis has reached a point at which extreme measures will have to be implemented. But who will control them, and in whose interests? Climate Capitalism offers a simple answer, namely, business as usual. Critical scholars like Lawrence Delina (2022:127) are rightly skeptical of 'risky Band-Aid approaches' like geoengineering, compared to 'demonstrated sustainable energy technologies.' The problem, however, is that under capitalist hegemony the crisis has advanced to a point of no return. Christian Parenti (2022:131) has posed the issue well:

> The science on climate tipping points is clear: even if we stopped all GHG pollution, we would need to strip CO_2 from the atmosphere. At the time of writing, CO_2 concentrations are 405 ppm (parts per million) and need to be at 350 ppm or lower to avoid self-compounding climate breakdown. In other words, stopping CO_2 emissions is not enough; we also need a global program of negative emissions.

'Negative emissions' necessarily implies geoengineering, to remove carbon already in the atmosphere. For Parenti, only a planned, coordinated, state-led response can accomplish this, but neoliberal market fetishism blocks such action. Parenti identifies a second blockage in 'the deep technophobia and nature fetish of many environmentalists.' In effect, this is the flipside to the techno-fetishism of ecological modernization. Environmentalists not grounded in historical materialism tend to have romantic views of nature as something external to us, which we must preserve. The two ideological blockages can be detected in the way those on the right and left engage with CDR as a negative-emissions technology.

> As a result of these ideological blockages, most of the people who support CDR technology operate with ridiculous ideas about market-based mechanisms for the technology's mass deployment. Meanwhile, those on the

> left and in environmental movements who think critically about capitalism and pressure the government to craft progressive policy remain largely silent on CDR or oppose it as just sophisticated technogreenwashing. Because CDR means large-scale technological intervention into the climate system, most greens reject it without further consideration. This is highly dangerous and wrongheaded.
>
> *(2022:131)*

Parenti points out that bringing CDR to scale as a profitable industry (i.e., the basic ecological modernization scenario) is a non-starter. The world could never make productive use of all the carbon to be removed and stored through CDR; this 'product' could never be marketed at profit. Like SRM (but for a different reason), CDR technology cannot be easily brought within capitalist social relations: its costs are too high and its benefits too diffuse. Thus, in Parenti's view, CDR should be taken up 'as a global technology commons and deployed by governments as a public utility' (2022:135). This means rejecting market fetishism. However, the state-led transformation will need to extend beyond a single techno-fix. 'A state-led crash program of CDR would only be meaningful within the context of a broader program of radical mitigation involving euthanizing the fossil fuel industry, a massive clean energy build-out, and robust adaptation efforts like coastal defense' (2022:139). Such a programme can only succeed through the collective agency of mass movements demanding radical, state-led transformation.

As this example intimates, the solutions offered by Climate Capitalism are false because of how they are positioned within a hegemonic project that refuses to address the elephant in the room: capital, its growth imperative and its anti-democratic power within human affairs. As Parenti reminds us, 'we face what Marx described as a contradiction between the forces of production and the relations of production' (2022:131). Climate Capitalism ignores this contradiction and focusses exclusively on transforming forces of production, in ways that suit the existing relations of production. But in fact, capitalism's social relations – the economic dominance of a small class of profit-driven capitalists, the financialized character of contemporary global accumulation, the imperative to 'grow or die' – are now holding back real possibilities for avoiding ecocide by transitioning to a fully sustainable and socially just way of life (Graham 2021).

Climate Capitalism, in its various renditions, amounts to what Gramsci called a 'passive revolution', a transformation 'from above' that is pushed along a conservative path that protects and even restores the basis for ruling-class power (Morton 2024:179). In Climate Capitalism 'A package of "green growth", corporate renewables, offsets for carbon sinks, "offshoring" of emissions, and "net zero" commitments are offered "from above"' (Goodman and Morton 2023:1260). Climate Capitalism strives to bring climate directly into the accumulation process, steering investment towards new green industries (Sapinski 2016). Its new ways of valorizing nature promise transformation 'without calling fundamental relations of power and domination into question' (Brand and Wissen 2021:177). Within the passive revolution, as we have seen, mechanisms for externalizing crises of the imperial mode of living gain acceptance 'as an immutable reality' (Brand and Wissen 2021:186). Studiously avoiding the capitalist relations of production, false solutions give license to renewed accumulation and eco-imperialism – shifting extractivism from carbon to lithium and other minerals, dispossessing Indigenous peoples to earn 'carbon credits' and to establish bioenergy plantations – all under the umbrella of 'clean growth.' Whereas stage 1 climate denialism was obstructionist, in passive revolution the ruling bloc embraces solutions that are false because they adhere, 'at every turn, to economic imperatives over other needs' (Buller 2022:23).

Those 'other needs' point to an ethically based reason why Climate Capitalism's solutions are false, which comes into view when we take a global perspective. In Brand and Wissen's take, ecological modernization aims to modernize and universalize the imperial mode of living, but in this initiative 'the global North is attempting to maintain something that cannot be maintained, and something that cannot exist on a universal basis is expanded and universalized in many countries of the global South.' In the face of 'growing upheaval and increasingly brutal externalizations' they flag another urgent need, 'for genuine alternatives that lead to a solidary mode of living, justice (both social and ecological), peace and democracy' (2021:187).

There is good reason to follow Brand and Wissen's lead, which I will do in the final chapter of this book. Carbon Tracker's recent analysis of global emissions reveals that in the five years from 2018 to 2023 the world exhausted more than one-third of its carbon budget (for a 50% chance of not exceeding a 1.5°C global temperature increase). Going forward, to stay within the 1.5°C limit, nearly 60% of the fossil fuels within extraction sites already operating or under-construction cannot be burned – meaning that most existing fossil-fuel infrastructure must be retired, an unlikely eventuality within Climate Capitalism (Coffin and Prince 2023). One might retort that renewable energy will provide the solution. But as Simon Michaux's (2021) detailed empirical analysis clearly shows, 'replacing the existing fossil fuel powered system (oil, gas, and coal), using renewable technologies, such as solar panels or wind turbines, will not be possible for the entire global human population.'[4] We are left, then, with the ethical issue flagged by Brand and Wissen. It is simply not possible to 'universalize' the current way of life to which people in the global North have become accustomed – post-Fordist consumer capitalism – while staying within well-established ecological boundaries. Maintaining the imperial mode of living within a deepening climate crisis, using the various technologies of ecological modernization, is a formula for ever-deepening climate apartheid. Humanity must shift to a way of life that is truly democratic and thus attuned to 'sustainable human development' (Magdoff and Foster 2011), not the narrow needs of capital for its own self-expansion, which currently, as 'growth', impersonate the general interest of humanity.

Finally, I must point out that the 'energy dilemma' between 'climate and growth' is increasingly viewed as a barrier to capital accumulation. As Cedric Durand (2021) notes, the mechanisms of Climate Capitalism are not enough to save the climate, but they are still 'proving too much for capitalist growth.' A 2023 study of 100 major oil and gas companies reveals that since 2021 the sector has made virtually no progress towards the Paris Agreement goals, as not one firm had cut emissions at a rate aligned with a 1.5°C pathway (Beer 2023). In March 2024, at a major conference of fossil capitalists and political managers in Houston, Amin Nasser, CEO of Saudi Aramco, stated, to applause, 'we should abandon the fantasy of phasing out oil and gas' – a sentiment 'backed up by record oil-and-gas production and consumer misgivings over purchasing electric vehicles' (Joselow 2024). While fossil capitalists dismissed a rapid energy transition (and have 'aggressively fought the Biden administration's efforts to boost sales of electric vehicles, which could eat into demand for their petroleum products') they acknowledged that their companies are benefitting from the Biden administration's signature climate law, which provides lucrative tax credits for CCUS (ibid.).

A recent study of investment plans by the world's six privately owned oil supermajors reports that

> the supermajors will continue a strategy of growing near-term oil and gas production whilst maintaining flexibility in their levels of low carbon spending, as they look to assess the speed of energy transition. ... None plan to cut production levels before 2025 and all plan to grow upstream by at least 2% per annum until then. Most speak openly about a strategy based on continued high fossil fuel demand, while BP and Shell last year rolled back targets on production cuts.
>
> *(Young 2024)*

The chasm between the 60% reduction in fossil extraction, needed to avoid runaway climate change, and the reality of capitalist investment plans foreshadows the future of Climate Capitalism. According to Emiliano Brancaccio (2023), a growing fraction of capitalists is challenging what they consider the excessive rigidity of measures to reduce carbon emissions. They view the 'green transition' as too fast-paced, risking the profits that drive 'growth' as production costs rise. He asks,

> ...why is it that the anti-ecological faction of capital is gaining ground just about everywhere, to the detriment of the one that is at least willing to flirt with environmental issues? The answer is as easy to give as it is bitter to acknowledge. Environmentally unfriendly capitalists are fishing for support among a fragmented working class, already battered by inflation, which, even though it might agree with the warnings about climate change, also seems increasingly impatient with the costs of the ecological transition.

Brancaccio wonders if any lessons can be drawn from the crisis in environmental politics. He answers that 'the ecological transition can only find acceptance among the masses if its social costs are not passed on to wages but rather to profits and rents, in a "law of environmental drop in the profit rate," a drop that would manage to head off the risk of climate catastrophe.' Yet falling profits are exactly what drives accumulation crisis. Even the magic of 'green finance' through the market-based mechanisms I have reviewed above cannot square the circle. Hence, 'however much it upsets the rich, whether they're pro-environment or not, there can only be one solution: a novel and innovative version of a collective plan' (ibid.). In this book's final chapter I will argue that democratic eco-socialism presents just such an alternative.

Notes

1 Among states with substantial populations (i.e., leaving aside small-population oil producers such as Qatar and Kuwait) as of 2022 Canada's per capita CO_2 emissions (14.2 tonnes per person) were somewhat below Saudi Arabia's (18.2 tonnes) and slightly below Australia's (15.0) and the United States' (14.9) (Global Carbon Budget 2023).

2 https://geoengineering.environment.harvard.edu/geoengineering accessed 26 February 2024.

3 An alternative to vast plantations is the development of 'mechanical trees', as is underway at the Arizona State University (Chalmin 2024). Such devices, obviously, offer no help in mitigating the rapid decline of biodiversity, which is a looming element of global ecological crisis, closely related to climate breakdown.

4 This is not an argument against developing renewable energy, whether from wind, solar, geothermal, tidal or other sources. In each case, the development process must be mindful of the socio-ecological relations involved in the entire lifecycle of energy production and usage. Healthy scepticism on all corporate and state-driven projects is a crucial aspect of this mindfulness, as scams abound. Currently, for instance, hydrogen, 'green' and even 'blue' (the latter created using natural gas, and thus a carbon emitter), is being touted by industry and government. A recent study shows that 'blue' hydrogen could produce 50% more global warming than burning fossil fuels directly (Beer 2024).

References

Adkin, Laurie, Mike Lang, Mark Shakespear, David Chen, and William K. Carroll. 2022. *Higher Education – Corporate or Public?* Vancouver: Canadian Centre for Policy Alternatives.

Albert, Michael J. 2021. "The Climate Crisis, Renewable Energy, and the Changing Landscape of Global Energy Politics." *Alternatives: Global, Local, Political* 46(3):89–98. doi: 10.1177/03043754211040698

Altvater, Elmer. 2016. "The Capitalocene, or, Geoengineering against Capitalism's Planetary Boundaries." Pp. 138–52 in *Anthropocene or Capitalocene? Nature, History, and the Crisis*, edited by Jason W. Moore. Oakland, CA: PM Press.

Bakx, Kyle. 2024. "$34B Trans Mountain Expansion Pipeline Begins Filling with Oil with First Shipments before Canada Day." *CBC News*. Retrieved April 2, 2024 (https://www.cbc.ca/news/canada/calgary/trans-mountain-expansion-begins-1.7150343).

Beer, Mitchell. 2023. "'Death Sentence for People, Ecosystems' as Fossil Companies Explore for More Oil." *The Energy Mix*. Retrieved March 1, 2024 (https://www.theenergymix.com/death-sentence-for-people-ecosystems-as-fossil-companies-explore-for-more-oil/).

Beer, Mitchell. 2024. "'Blue' Hydrogen Could Produce 50% More Warming than Burning Fossil Fuels." *The Energy Mix*. Retrieved March 1, 2024 (https://www.theenergymix.com/blue-hydrogen-could-produce-50-more-warming-than-burning-fossil-fuels/).

Belot, Ross. 2024. "We Have Already Reached Post-Peak Oil." *National Observer*. Retrieved (https://www.nationalobserver.com/2024/02/02/opinion/demand-oil-market-Alberta-demand-energy-transition).

Betts, Alan K. 2021. "Climate Change and Society." *AIMS Geosciences* 7(2):194–218. doi: 10.3934/geosci.2021012

Brancaccio, Emiliano. 2023. "Why Don't Capitalists Want to 'Go Green' Anymore?" *The Bullet*. Retrieved February 29, 2024 (https://socialistproject.ca/2023/10/why-dont-capitalists-want-to-go-green-anymore/).

Brand, Ulrich, and Markus Wissen. 2021. *The Imperial Mode of Living*. London: Verso.

Brulle, Robert J. 2018. "The Climate Lobby: A Sectoral Analysis of Lobbying Spending on Climate Change in the USA, 2000 to 2016." *Climatic Change* 149(3–4):289–303. doi: 10.1007/s10584-018-2241-z

Buch-Hansen, Hubert, and Peter Nielsen. 2023. "Critical Realism, the Climate Crisis and (de)Growth." *Journal of Critical Realism* 22(3):347–63. doi: 10.1080/14767430.2023.2217050

Buck, Holly Jean, J. P. Sapinski, and Andreas Malm. 2022. "Critical Perspectives on Geoengineering: A Dialogue." Pp. 3–20 in *Has It Come to This? The Promises and Perils of Geoengineering on the Brink*, edited by J. P. Sapinski, H. J. Buck, and A. Malm. New Brunswick, New Jersey: Rutgers University Press.

Buller, Adrienne. 2022. *The Value of a Whale: On the Illusions of Green Capitalism*. Manchester: Manchester University Press.

Bumpus, Adam G., and Diana M. Liverman. 2008. "Accumulation by Decarbonization and the Governance of Carbon Offsets." *Economic Geography* 84(2):127–55. doi: 10.1111/j.1944-8287.2008.tb00401.x

Canada. 2024. "Greenhouse Gas Emissions." *Environment and Climate Change Canada*. Retrieved May 29, 2024 (www.canada.ca/en/environment-climate-change/services/environmental-indicators/greenhouse-gas-emissions.html.).

Carroll, William K. 2017. "Canada's Carbon-Capital Elite: A Tangled Web of Corporate Power." *Canadian Journal of Sociology* 42:225–60.

Carroll, William K. 2022. "Marx Matters, in Theory and Practice: Reflections from the Corporate Mapping Project." Pp. 198–225 in *Marx Matters*, edited by D. Fasenfest. Leiden: Brill.

Carroll, William K. 2021. "Conclusion: Prospects for Energy Democracy in the Face of Passive Revolution." Pp. 479–504 in *Regime of Obstruction*, edited by W. K. Carroll. Edmonton: AU Press.

Carroll, William K., Shannon Daub, and Shane Gunster. 2022. "Regime of Obstruction: Fossil Capitalism and the Many Facets of Climate Denial in Canada." Pp. 216–33 in *Handbook of Anti-Environmentalism*, edited by David Tindall, Mark C.J. Stoddart and Riley E. Dunlap. Cheltenham, UK: Edward Elgar Publishing.

Carroll, William K., Nicolas Graham, Michael Lang, Kevin McCartney, and Zoe Yunker. 2021. "Fossil Capital's Reach into Civil Society: The Architecture of Climate Change Denialism." Pp. 171–96 in *Regime of Obstruction: How Corporate Power Blocks Energy Democracy*, edited by William K. Carroll. Edmonton: AU Press.

Carroll, William K., Nicolas Graham, and Mark Shakespear. 2020. "Foundations, ENGOs, Clean Growth Networks and the Integral State." *Canadian Journal of Sociology* 45(2):109–42.

Carroll, William K., and Jouke Huijzer. 2021. "Who Owns Big Carbon? Mapping the Network of Corporate Ownership." Pp. 111–41 in *Regime of Obstruction: How Corporate Power Blocks Energy Democracy*, edited by William K. Carroll. Edmonton: AU Press.

Carton, Wim. 2019. "'Fixing' Climate Change by Mortgaging the Future: Negative Emissions, Spatiotemporal Fixes, and the Political Economy of Delay." *Antipode* 51(3):750–69. doi: 10.1111/anti.12532

Castree, Noel, and Brett Christophers. 2015. "Banking Spatially on the Future: Capital Switching, Infrastructure, and the Ecological Fix." *Annals of the Association of American Geographers* 105(2):378–86. doi: 10.1080/00045608.2014.985622

Cecco, Leyland. 2023. "Wildfires Turn Canada's Vast Forests from Carbon Sink into Super-Emitter." *The Guardian*. Retrieved February 26, 2024 (https://www.theguardian.com/world/2023/sep/22/canada-wildfires-forests-carbon-emissions).

Chambers, Collin. 2021. "A Critique of the 'Socio-ecological Fix' and towards Revolutionary Rupture." *Area* 51:114–121.

Chalmin, Anja. 2024. "They Can't See the Wood for the MechanicalTrees™: Recent Direct Air Capture Projects Are More Successful at Capturing Funding than Carbon." *Geoengineering Monitor*. Retrived August 28, 2024 (https://www.geoengineeringmonitor.org/2024/02/geo-map-dac-feb-2024/).

Clark, Brett, and Richard York. 2005. "Carbon Metabolism: Global Capitalism, Climate Change, and the Biospheric Rift." *Theory and Society* 34(4):391–428.

Coffin, Mike, and Guy Prince. 2023. *Navigating Peak Demand*. London: Carbon Tracker Initiative. Retrieved August 28, 2024 (https://carbontracker.org/reports/navigating-peak-demand/).

Cran-McGreehin, Simon. 2021. "Negative Emissions: Why, What, How?" *Energy and Climate Intelligence Unit*. Retrieved February 24, 2024 (https://eciu.net/analysis/briefings/net-zero/negative-emissions-why-what-how).

Daggett, Cara. 2018. "Petro-Masculinity: Fossil Fuels and Authoritarian Desire." *Millennium: Journal of International Studies* 47(1):25–44. doi: 10.1177/0305829818775817

Delina, Lawrence L. 2022. "Mobilizing in a Climate Shock: Geoengineering or Accelerated Energy Transition?" Pp. 121–29 in *Has It Come to This? The Promises and Perils of Geoengineering on the Brink*, edited by J. P. Sapinski, H. J. Buck, and A. Malm. New Brunswick, New Jersey: Rutgers University Press.

Derber, Charles. 2010. *Greed to Green*. London: Paradigm Publishers.

DeSmith, Christy. 2023. "It May Sound Counterintuitive but You Probably Don't Drive Enough, Says Grad's Research." *The Harvard Gazette*. Retrieved (https://news.harvard.edu/gazette/story/2023/08/when-buying-an-ev-increases-your-carbon-footprint/).

Dockstader, Sue, and Shannon Elizabeth Bell. 2020. "Ecomodern Masculinity, Energy Security, and Green Consumerism: The Rise of Biofuels in the United States." *Critical Sociology* 46(4–5):643–60. doi: 10.1177/0896920519885010

Dunlop, R. E., and A. M. McCright. 2015. "Challenging Climate Change: The Denial Countermovement." Pp. 300–332 in *Climate Change and Society: Sociological Perspectives*, edited by R. Brulle and R. E. Dunlop. New York: Oxford University Press.

Durand, Cedric. 2021. "Energy Dilemma." *New Left Review – Sidecar*. Retrieved February 29, 2024 (https://newleftreview.org/sidecar/posts/energy-dilemma).

Dyke, James, Robert Watson, and Wolfgang Knorr. 2021. "Climate Scientists: Concept of Net Zero Is a Dangerous Trap." *The Conversation*. Retrieved February 28, 2024 (https://theconversation.com/climate-scientists-concept-of-net-zero-is-a-dangerous-trap-157368).

Eaton, Emily, and Simon Enoch. 2021. "The Oil Industry Is Us: Hegemonic Community Economic Identity in Saskatchewan's Oil Patch." Pp. 307–30 in *Regime of Obstruction*, edited by W. K. Carroll. Edmonton: AU Press.

Energy Institute. 2024. "Electricity." *Energy Institute*. Retrieved April 16, 2024 (https://www.energyinst.org/exploring-energy/topic/electricity).

Evans, Simon. 2023. "Factcheck: 21 Misleading Myths about Electric Vehicles." *CarbonBrief*. Retrieved August 28, 2024 (https://www.carbonbrief.org/factcheck-21-misleading-myths-about-electric-vehicles/).

Fazi, Thomas. 2024. "The Delusions of the Climate Power Bloc: Solar Geoengineering Is Just Another Fantasy." *UnHerd*. Retrieved August 28, 2024 (https://unherd.com/2024/02/the-delusions-of-the-climate-power-bloc/).

Foster, John Bellamy. 2010. "Why Ecological Revolution?" *Monthly Review* 61(8):1. doi: 10.14452/MR-061-08-2010-01_1

Fressoz, Jean-Baptiste, and Christophe Bonneuil. 2017. "Growth Unlimited: The Idea of Infinite Growth from Fossil Capitalism to Green Capitalism." Pp. 52–68 in *History of the Future of Economic Growth: Historical Roots of Current Debates on Sustainable Degrowth*, edited by I. Borowy and M. Schmelzer. London; New York, NY: Routledge, Taylor & Francis Group.

Friedemann, Alice J. 2021. *Life After Fossil Fuels*. Cham, Switzerland: Springer Nature.

Gelderloos, Peter, and Alexander Dunlap. 2023. "'The Poisons Are Already In Here with Us:' Framing for Ecological Revolutions from Below." *Globalizations* 1–18. doi: 10.1080/14747731.2023.2225306

Ghosh, Jayati, Shouvik Chakraborty, and Debamanyu Das. 2022. "Climate Imperialism in the Twenty-First Century." *Monthly Review* 70–85. doi: 10.14452/MR-074-03-2022-07_4

Global Carbon Budget. 2023. "Population based on Various Sources." *This World in Data*. Retrieved August 28, 2024. (https://ourworldindata.org/co2-and-greenhouse-gas-emissions?insight=there-are-large-differences-in-emissions-across-the-world#key-insights).

Goodman, James, and Tom Morton. 2023. "'Rage into Action': From Carbon Democracy to Climate Democracy?" *Globalizations* 20(8):1257–76. doi: 10.1080/14747731.2023.2270193

Graham, Nicolas. 2021. *Forces of Production, Climate Change and Canadian Fossil Capitalism*. Leiden: Brill.

Graham, Nicolas, William K. Carroll, and David Chen. 2020. "Carbon Capital's Political Reach: A Network Analysis Of Federal Lobbying By The Fossil Fuel Industry From Harper To Trudeau." *Canadian Political Science Review* 14(1):1–31. doi: 10.24124/c677/20201743

Gramsci, Antonio. 1971. *Selections from the Prison Notebooks*. edited by Q. Hoare and G. Nowell-Smith. New York: Lawrence & Wishart.

Griffin, Paul. 2017. *The Carbon Majors Database*. Snowmass, CO: Climate Accountability Institute.

Gunderson, Ryan, Diana Stuart and Brian Petersen, 2018 "Ideological Obstacles to Effective Climate Policy." *Capital & Class* 42(1):133–160.

Huber, Matthew T. 2013. *Lifeblood: Oil, Freedom, and the Forces of Capital*. Minneapolis: University of Minnesota Press.

Hudson, Mark. 2021. "Landscapes of Risk: Financial Representations of Catastrophe." Pp. 91–110 in *Regime of Obstruction: How Corporate Power blocks Energy Democracy*, edited by W. K. Carroll. Edmonton: AU Press.

Hudson, Mark, and Evan Bowness. 2021. "Finance and Fossil Capital: A Community Divided?" *The Extractive Industries and Society* 8(1):383–94. doi: 10.1016/j.exis.2020.11.016

Hughes, J. D. 2024. "Getting to Net-Zero in Canada: Summary." *Policynote*. Retrieved April 3, 2024 (https://www.policynote.ca/net-zero-2/).

Ilc, Blaž Vrečko. 2021. "Climate Crisis: Time to Rethink Economic Planning by Demistifying Capitalism and its Market(s)." *Teorija in praksa* 341–60. doi: 10.51936/tip.58.2.341-360

Intergovernmental Panel On Climate Change (IPCC). 2023. *Climate Change 2022 – Impacts, Adaptation and Vulnerability: Working Group II Contribution to the Sixth Assessment Report of the Intergovernmental Panel on Climate Change*. 1st ed. Cambridge, UK: Cambridge University Press.

International Energy Agency. 2023. *CO_2 Emissions in 2022*. Retrieved September 17, 2024 (https://www.iea.org/reports/co2-emissions-in-2022).

Joselow, Maxine. 2024. "Phasing out Fossil Fuels a 'Fantasy,' Oil Executives Say amid Giant Profits." *Washington Post*. Retrieved March 20, 2024 (https://www.washingtonpost.com/climate-environment/2024/03/20/oil-industry-climate-fossil-fuels/).

Lachapelle, Paul, Patrick Belmont, Marco Grasso, Roslynn McCann, Dawn H. Gouge, Jerri Husch, Cheryl De Boer, Daniela Molzbichler, and Sarah Klain. 2024. "Academic Capture in the Anthropocene: A Framework to Assess Climate Action in Higher Education." *Climatic Change* 177(3):40. doi: 10.1007/s10584-024-03696-4

Lang, Chris. 2024. "Voluntary Carbon Markets: 'Decades of Misrepresentation, Mismanagement, and Fraud.'" *REDD-Monitor*. Retrieved February 26, 2024 (https://reddmonitor.substack.com/p/voluntary-carbon-markets-decades).

Laville, S. 2019. "Top Oil Firms Spending Millions Lobbying to Block Climate Change Policies, Says Report." *The Guardian*. Retrieved August 28, 2024 (https://www.theguardian.com/business/2019/mar/22/top-oil-firms-spending-millions-lobbying-to-block-climate-change-policies-says-report).

Legault, Frederic, Arnaud Theurillat-Cloutier, and Alain Savard. 2023. *Ecology for the 99%*. Toronto: Between the Lines.

Levin, Julia. 2024. "Canada's Fossil Fuel Funding, 2023." *Environmental Defence*. Retrieved April 17, 2024 (https://environmentaldefence.ca/wp-content/uploads/2024/03/Canadas-Fossil-Fuel-Subsidies.pdf).

Lopez, Johannah. 2023. "Algae to Save the World?" *DevX*. Retrieved February 28, 2024 (https://www.devx.com/news/algae-to-save-the-world/).

Lourie, Bruce. 2024. "Canada Can Thrive and Achieve Climate Success with or without a Carbon Tax." *National Observer*. Retrieved September 17, 2024 (https://www.nationalobserver.com/2024/03/08/opinion/canada-can-achieve-climate-success-without-carbon-tax).

Magdoff, Fred, and John Bellamy Foster. 2011. *What Every Environmentalist Needs to Know about Capitalism*. New York: Monthly Review Press.

Malm, Andreas. 2022. "Planning the Planet: Geoengineering Our Way out of and Back into a Planned Economy." Pp. 143–62 in *Has it come to This? The Promises and Perils of Geoengineering on the Brink*. New Brunswick, NJ: Rutgers University Press.

Malm, Andreas, and Wim Carton. 2021. "Seize the Means of Carbon Removal: The Political Economy of Direct Air Capture." *Historical Materialism* 29(1):3–48. doi: 10.1163/1569206X-29012021

McCreary, Tyler. 2021. "Crisis in the Tar Sands: Fossil Capitalism and the Future of the Alberta Hydrocarbon Economy." *Historical Materialism* 30(1):31–65. doi: 10.1163/1569206X-12341971

Michaux, Simon P. 2021. *Assessment of the Extra Capacity Required of Alternative Energy Electrical Power Systems to Completely Replace Fossil Fuels*. Espoo, Finland: Geological Survey of Finland.

Misra, Siddharth. 2023. "Plummeting 'Energy Return on Investment' of Oil and the Impact on Global Energy Landscape." *Journal of Petroleum Technology*. Retrieved February 23, 2024 (https://jpt.spe.org/plummeting-energy-return-on-investment-of-oil-and-the-impact-on-global-energy-landscape).

Mol, Arthur P. J. 1995. *The Refinement of Production*. Utrecht: International Books.

Morton, Adam David. 2024. "Where Trotsky's Horizons Stop, Gramsci's Begin: The Passive Revolutionary Road to Capitalist Modernity." Pp. 171–88 in *The Elgar Compnion to Antonio Gramsci*, edited by William K. Carroll. Cheltenham, UK: Edward Elgar.

Neubauer, Robert, Nicolas Graham, and Helena Krobath. 2023. "Defending 'Canadian Energy': Connective Leadership and Extractive Populism on Canadian Facebook." *Environmental Communication* 17(6):634–52. doi: 10.1080/17524032.2023.2235919

Nuwer, Rachel. 2013. "Oil Sands Mining Uses Up Almost as Much Energy as It Produces." *Inside Climate News*. Retrieved February 23, 2024 (https://insideclimatenews.org/news/19022013/oil-sands-mining-tar-sands-alberta-canada-energy-return-on-investment-eroi-natural-gas-in-situ-dilbit-bitumen/).

Oil Change International. 2023. *Banking on Climate Chaos 2023: Fossil Fuel Finance Report*. Oil Change International. Retrieved August 28, 2024. (https://priceofoil.org/2023/04/13/banking-on-climate-chaos-2023/)

Parenti, Christian. 2022. "A Left Defense of Carbon Dioxide Removal: The State Must Be Forced to Deploy Civilization-Saving Technology." Pp. 130–42 in *Has it Come to This? The Promises and Perils of Geoengineering on the Brink*. New Brunswick, NJ: Rutgers University Press.

Pineault, Éric. 2023. *A Social Ecology of Capital*. London: Pluto Press.

Public Citizen. 2024. "Public Citizen Comment on CFTC's Proposed Carbon Offsets Derivatives Products." Retrieved February 26, 2024 (https://www.citizen.org/article/cftc-proposed-carbon-offsets-derivatives/).

Rathi, Akshat. 2024. *Climate Capitalism: Winning the Race to Zero Emissions and the Crisis of Our Age*. Vancouver: Greystone Books.

Richards, Carol, and Kristen Lyons. 2016. "The New Corporate Enclosures: Plantation Forestry, Carbon Markets and The Limits of Financialised Solutions to The Climate Crisis." *Land Use Policy* 56:209–16. doi: 10.1016/j.landusepol.2016.05.013

Ritchie, Hannah. 2021. "A Number of Countries Have Decoupled Economic Growth from Energy Use, Even If We Take Offshored Production into Account." *Our World In Data*. Retrieved August 28, 2024 (https://ourworldindata.org/energy-gdp-decoupling).

Sapinski, Jean Philippe. 2016. "Constructing Climate Capitalism: Corporate Power and the Global Climate Policy-Planning Network." *Global Networks* 16(1):89–111.

Singh, Siddharth. 2024. "The Relationship between Growth in GDP and CO2 Has Loosened; It Needs to Be Cut Completely." *The Energy Mix*. Retrieved May 21, 2024 (https://www.iea.org/commentaries/the-relationship-between-growth-in-gdp-and-co2-has-loosened-it-needs-to-be-cut-completely).

Smil, Vaclav. 2022. *How the World Really Works*. London: Penguin Books.

Steward, Gillian. 2017. *Betting on Bitumen: Alberta's Energy Policies from Lougheed to Klein*. Vancouver: Canadian Centre for Policy Alternatives.

Stuart, Diana, Ryan Gunderson, and Brian Petersen. 2020. "The Climate Crisis as a Catalyst for Emancipatory Transformation: An Examination of the Possible." *International Sociology* 35(4):433–56. doi: 10.1177/0268580920915067

Surprise, Kevin. 2018. "Preempting the Second Contradiction: Solar Geoengineering as Spatiotemporal Fix." *Annals of the American Association of Geographers* 108(5):1228–44. doi: 10.1080/24694452.2018.1426435

Surprise, Kevin. 2024. "Beyond Ecocidal Capitalism: Climate Crisis and Climate Justice." Pp. 448–68 in *The Elgar Companion to Antonio Gramsci*, edited by W. K. Carroll. Cheltenham, UK: Edward Elgar.

Surprise, Kevin, and Jean Philippe Sapinski. 2023. "Whose Climate Intervention? Solar Geoengineering, Fractions of Capital, and Hegemonic Strategy." *Capital & Class* 47(4):539–64. doi: 10.1177/03098168221114386

Tasker, John Paul. 2022. "Canada Releases Plan for a 40 per Cent Cut in Carbon Emissions by 2030." *CBC*. Retrieved April 3, 2024 (https://www.cbc.ca/news/politics/canada-2030-emissions-reduction-plan-1.6401228).

Young, Tom. 2024. "Unpacking Energy Supermajors' Future Growth and Investment Strategies." *Evaluate Energy*. Retrieved February 29, 2024 (https://blog.evaluateenergy.com/insight-unpacking-energy-oil-gas-supermajors-future-growth-and-low-carbon-esg-investment-strategies).

Vitello, Connie. 2019. "Canadians Demonstrate Climate Action Leadership at COP25." *Environment Journal*. Retrieved April 17, 2024 (https://environmentjournal.ca/canada-demonstrates-climate-action-leadership-at-cop25/).

Wood, Johnny. 2021. "What Does Net-Zero Emissions Mean and How Can We Get There?" *World Economic Forum*. Retrieved February 24, 2024 (https://www.weforum.org/agenda/2021/11/net-zero-emissions-cop26-climate-change/).

Woodside, John. 2024. "Carbon Capture and Hydrogen Tax Credits Expected to Cost the Government over $11 Billion: PBO." *National Observer*. Retrieved April 17, 2024 (https://www.nationalobserver.com/2024/02/02/news/carbon-capture-hydrogen-tax-credits-expected-cost-over-11-billion).

Wright, Christopher, and Daniel Nyberg. 2015. *Climate Change, Capitalism, and Corporations*. Cambridge: Cambridge University Press.

5

ALTERNATIVES TO FOSSIL CAPITALISM?

Fortunately, the false solutions reviewed in Chapter 4 are not the only initiatives on offer in addressing the climate crisis. This chapter reviews several well-established alternatives to the status quo that might move us towards real solutions. In thinking about them, it is useful again to invoke the concept of project. We saw in previous chapters how Fossil Capitalism, emerging in the industrial revolution, gained hegemony in an era of competitive capitalism and consolidated its hegemony in the early decades of the 20th century, within a framework of 'carbon democracy' in the global North and eco-imperialism in the global South. The hegemonic project of post- World War 2 Fordism during the three boom decades (mid-1940s to mid-1970s) further strengthened that hegemony within a framework of class compromise and consumer capitalism, as carbon emissions sharply accelerated. That project came up against its own contradictions and was supplanted by transnational neoliberalism and Global post-Fordism. Yet, after the 2008 financial meltdown, neoliberal post-Fordism was hoisted by its own petard, although it continues to stumble along without a clear alternative from on high or below. As the civilizational crisis deepens, carbon pollution exacerbates climate breakdown, now increasingly visible around the world, from the melting of polar icecaps to extreme weather events and desertification in temperate and tropical climate zones. What projects can provide an escape hatch from looming ecocide?

In their 2018 book, *Climate Leviathan*, Wainwright and Mann made a useful contribution towards answering this question. They employed a Weberian ideal-typical analysis to highlight four possible projects in our era of deepening climate crisis. Each project 'is distinguished by the hegemony of a particular bloc, a mode of appropriation and distribution through which

DOI: 10.4324/9781003412922-8

that hegemony is exercised: a capitalist Climate Leviathan; an anticapitalist, state-centered Climate Mao; a reactionary capitalist Behemoth; and an anti-capitalist, anti-sovereign Climate X' (Wainwright and Mann 2018:29). Climate Leviathan represents the elite, Northern-driven project of Climate Capitalism, including both decarbonization through ecological modernization and planetary-level, sovereign control of climate policy, as in the 'central planning' required for geoengineering. Climate Mao points towards a state-led break from capitalism and endless growth, led by China (which constitutionally embraced a project of Ecological Civilization in 2018; Hansen et al. 2018) in a broader global South configuration. Behemoth foresees the continuation of Fossil Capitalism, as populist, particularly fascist, forces reject the notion of a planetary sovereign, sticking instead with nationally organized state power and market anarchy, protected by shoring up borders against a rising tide of climate refugees (also sea levels!). If Behemoth is a dystopic projection personified by real-life figures like Donald Trump and Jair Bolsonaro, the fourth project, Climate X, offers an anarchist utopian vision of post-capitalism that breaks from both capitalist and state power.

Wainwright and Mann's book has had influence in academe and in movements (e.g., Sen 2019; Asher 2020; Levrat 2020), but in my view, they are led astray by the Weberian method of ideal types, which purposefully accentuates certain features of reality, creating abstract, relatively static scenarios that lack dialectical sensibilities. In particular, the rhetorical move to keep the state-centred 'Climate Mao' sealed off from anti-statist 'Climate X' reproduces a socialism vs anarchism binary (with the authors clearly preferring the latter on ethical grounds while providing no reasoning as to its viability). As I will argue in the final chapter, a robust, eco-socialist project, which proceeds from a reformulation of 'planning', can undo the binary that these authors create for themselves using the Weberian method.

This chapter is organized in three parts. First, I lay out the challenges and barriers to change in the current era, and the need to go beyond 'business of politics as usual', including social-democratic incrementalism. The situation in the global North is the focus of these reflections. In a political culture still dominated by neoliberalism and hostile to socialism, projects opposing fossil capital have been framed in two main ways in the global North: as green new deal reformism (including just transition programmes) and as degrowth. These North-based projects have the advantage of resonating, in

TABLE 5.1 Four potential social formations vis-a-vis the climate crisis

	Planetary sovereignty	*Anti-planetary sovereignty*
Capitalist	Climate Leviathan	Climate Behemoth
Non-capitalist	Climate Mao	Climate X

Source: Wainwright and Mann (2012:5).

different ways, with immediate popular concerns and common sense. They have the disadvantage of obscuring the deeper causes of the problem, the causal power of capital itself. Within these projects, the predominant political current fits comfortably within liberal ideology. Yet currents also active within and around them have transformative potential, if they can be articulated with a more comprehensive project. A third project, emanating from Indigenous lifeways in the global South (specifically, the Andean region of South America) is Buen Vivir – 'living well' – which presents a deep critique of imperialism and extractivist capitalism. Each of these initiatives moves in the right direction, but none of them offers the comprehensive approach of eco-socialism, which is the focus of this book's final chapter.

Challenges and Barriers

The barriers to escaping our collective demise are both structural and cultural/psychological. The structure of Fossil Capitalism and imperialism puts in place, and keeps in place, a bloc of capitalists and their organic intellectuals, who govern the state and economy. Subalterns are positioned largely as passive recipients of decisions made over their heads. They may resist domination, but, unless they are organized behind a counter-hegemonic project, their resistance tends to be fragmented and episodic. Moreover, to repeat a point from Chapter 2, the structure of market society has an atomizing impact on the class that, potentially, could transform that structure. For the atomized worker, 'all other workers are competitors; all other workers are enemies insofar as they are competing for the same jobs' (Lebowitz 2020:111).

These basic structural features lead us immediately to the cultural/psychological terrain, extending from the common sense of everyday life to well-formed ideologies identified with hegemonic struggle. Here, several barriers specific to effective climate action appear, both in the lived reality of individuals and in the collective, political culture of neoliberal capitalism. Beginning with the former, social scientists investigating emotional life in the context of climate crisis have documented a number of barriers to climate action, which amplify the psycho-cultural tendencies already ingrained in the era of Fordist consumer capitalism – the fetish of growth, the privatism of automobility and suburbia etc. (see Chapter 2). It is worthwhile to review the main barriers.

A now-classic ethnography of climate denialism, Kari Norgaard's study of a community in northern Norway revealed how 'socially organized denial' normalized a way of being in which caring about ecological conditions is 'actively muted in order to protect individual identity and sense of empowerment and to maintain culturally produced conceptions of reality' (2011:207). She showed how, through 'emotion management strategies' like selective attention, 'problematic' emotions of fear and helplessness were marginalized while normative emotions of being optimistic and maintaining control of one's life retained their salience (2011:174). Norgaard's research led her to conclude that

> people create a sense of everyday reality using features of everyday life from emotion norms to cultural narratives. Thus, denial is 'socially organized' not only because we think as members of groups, but because societies develop and reinforce a whole repertoire of techniques or 'tools' for ignoring disturbing problems.
>
> *(2011:215)*

Norgaard's book has inspired a great deal of further inquiry into the sociology of climate change denial (Tindall et al. 2022). But climate inaction also stems from a pervasive sense of powerlessness among the atomized subjects of market society. Using the United States as exemplar, Ryan Gunderson argues that an objective structure of 'real helplessness', taken up as a lived reality by atomized individuals, presents a barrier between climate concern and action. For Gunderson, real helplessness has three dimensions: powerlessness (in the face of entrenched capitalist social relations coupled with the absence of social organizations with the power to challenge the status quo), stupefaction (the power of the culture industry) and repression (a state apparatus that 'uses violence to squash any militant and organized opposition') (2023:7). These political-economic conditions, highly developed in the United States but endemic across the late capitalist world, 'close off avenues for performing actions that are consistent with concern for the environment' (2023:18). In a self-fulfilling prophecy, the lack of a political infrastructure to enact an alternative to capitalism, and the threat of state repression, lead many to a politically passive, 'realist' stance, prompting Gunderson to ask, 'how can helpless people transition out of a social formation that is even resistant to reforms?' (2023:19). Similarly, Audrea Lim (2018), reflecting on the persuasive power of the 'ideology of fossil fuels' – 'the freedom to express one's identity through mass, conspicuous consumption' – suggests that the enormity of the climate crisis has engendered a widespread environmental dread that fuels stress and depression, pacifying the people in the face of unrelenting petrocapitalism.

This is not to say that there is no climate action from below. However, actions often replicate a tendency in mainstream environmentalism towards 'depoliticization' (Swyngedouw 2013), as in acts of 'ecopiety' to atone for our sins while our everyday lives 'perpetuate the logics of global capitalism and market ideology' (Taylor 2019:3). Such approaches orient people towards individualized acts of 'green consumerism', towards 'feelings of sadness and guilt that make us want to quickly expiate our guilt and feel better', and away from 'anger at the corporations and government institutions that are responsible for the systems in which we act' (Vaughan 2021:277). Reinforcing all these barriers to real climate action is a 'statephobia' that in the era of neoliberal hegemony has become influential on the left, leaving many activists unable 'to see the state and comprehend its central role in modern life' (Parenti 2022:135).

In these psycho-cultural barriers to effective climate action, we find the other pole of Climate Capitalism as a multi-scalar passive revolution, working at levels ranging from the global (COP negotiations, carbon markets, geo-engineering, etc.), through the 'clean growth' policies of national states, right down to the everyday conversations, desires and purchases of us all. As I have stated elsewhere, Climate Capitalism's system-friendly reforms are a formula for elite-managed 'continuity in change', appealing 'to subjectivities already normalized within fossil-fuelled consumer capitalism' and portending no more than minor shifts in capitalism's historical bloc (Carroll 2021:486). Within the bloc, the hegemonic project of Climate Capitalism as a policy paradigm shapes the actual lives of subalterns whose consent is recruited by those policies. Yet as Huber (2023) points out, the paradigm's 'techno-behavioural' approach, insisting that 'rather than winning them over to an attractive political project, the masses must be reformed into more virtuous low-carbon practices,' has not won hearts and minds. Instead, increasingly coercive measures to stamp out the 'irresponsible' choices of millions of consumers, to get them to drive less, eat less meat, fly less, etc., have spurred populist reaction, driving many to the extractive populism I discussed in the previous chapter.

The struggle for hegemony is a struggle for the future. Today, although fossil fuels continue to power most economic activities, the hegemonic struggle is between a Climate Capitalism led by powerful capitalist fractions and their organic intellectuals, and a nascent radical alternative that breaks from capitalism, in favour of caring for ourselves and for nonhuman nature. Climate Capitalism is increasingly preferred at the top of the global class structure, although corporations heavily committed to fossil fuels – as producers or consumers – will continue to pursue their immediate interests even as the energy transition rolls out. As a passive revolution, Climate Capitalism operates over the heads of the masses, which as we see, presents challenges in recruiting energetic support. Those challenges offer opportunities to a counter-hegemonic project intent on creating a post-capitalist, ecologically healthy future. To counter a passive revolution, Gramscian analysts have recommended the conduct of an 'anti-passive revolution': a war of position that extends popular-democratic and class struggles, 'to mobilize ever-wider sections of the population for democratic reforms' (Simon 1982: 49; Buci-Glucksmann 1979). In the chapter following this one, I will explicate how eco-socialism gives us the best hope for a comprehensive alternative. For now, as a prelude to discussing approaches that fall short of the eco-socialist alternative but that contain important elements convergent with that project,

I want to take up a barrier sometimes overlooked by critics of contemporary capitalism: the persuasive power of social democracy as the longstanding alternative, *within capitalism*, to (neo)liberalism.

Social democracy emerged in Europe late in the 19th century, as the growing power of organized labour appeared to open a strategy of incrementally reforming capitalism into a way of life amenable to workers. Leaders like Eduard Bernstein called for 'evolutionary socialism', rejecting a vision of revolution with the phrase, 'the ultimate aim of socialism is nothing, but the movement is everything' (Bernstein 1899). In World War 1, a key difference between revolutionary socialism and social democracy crystallized, as social democratic parties embraced nationalism and supported their respective warring states while revolutionary socialists like Lenin criticized imperialism and called on proletarian soldiers to train their weapons on the source of their oppression – the ruling class. Out of that epic conflict came both state socialism and social-democratic reformism. The latter became a predominant political current, as we saw in Chapter 2, during the post- World War 2 era, as workers in the global North made major gains within the class compromise of carbon democracy, which also tied them to the fortunes of their national capitalist classes and states.

As a project, social democracy assumed that through state management of the market economy, capitalist crises could be tempered, if not avoided. Instead of seizing and transforming the state, 'social democracy could continue to grow in strength, extract piecemeal reforms and gradually lift the working class out of the mire' (Malm 2020:74). Capitalism's deep crises – in the 1930s, the 1970s and in 2008 onward – derailed this assumption, as they incubated anti-democratic far-right movements: fascism, neoliberalism and today's extractive populism (extending to neo-fascism). In each case, the social democratic left was stymied by an organized right intent on preserving a market-based way of life dominated by capital. Meanwhile, as we have seen, in the era of class compromise, social democracy's heyday, reforms co-opted Northern workers into the imperial mode of living:

> the Left gradually came to abandon the original qualitative notion of a democratic society (socialism) and adopt a quantitative understanding of democracy – as equality of citizenship within a capitalist society. This version of democracy is not only perfectly compatible with the dynamics of competitive profit production, but is dependent on it – we need to grow the pie in order to distribute it – with all the attendant dynamics of exploitation, alienation and destruction of nature.
>
> *(Azmanova 2019:1197)*

Social democracy's defeat in the 1980s signalled that the globalization and deregulation of capital had greatly narrowed the scope for nationally based reform (Teeple 2008). Transnationalizing capitalists could, threatening disinvestment, play one state off against another, pushing all states into a race to the bottom. Thereafter, social democratic parties themselves shifted to the right, pragmatically adopting many of neoliberalism's priorities. The aftermath of 2008 has also not been kind to social democracy. Most social-democratic parties now inhabit the margins of political life. But the neoliberal project is also threadbare, its protagonists increasingly opting for authoritarian measures (Bruff 2014). In Europe, former Greek Finance Minister Yanis Varoufakis (2024) has concluded that 'democracy is finished.' Two anti-democratic and co-dependent currents now dominate the political spectrum: the authoritarian neoliberal establishment and the neofascist right that has formed in part as a reaction to neoliberalism's brutality and social democracy's impotence. A similar narrative could be constructed for capitalist democracies elsewhere.

Within the parameters of contemporary capitalism, democracy's future seems in doubt. Perhaps, as Albena Azmanova (2019:1197) suggests, '"democratic capitalism" is an oxymoron: the goals of a democratic society – one committed to collective goals – are by definition incompatible with capitalism's constitutive dynamic (and ergo, key interest) – the perpetuation of capital accumulation.' For social democracy, the rapidly closing window for collective action that could avoid climate catastrophe poses a further challenge. If the (democratic) *space* for social democracy has shrunk, so has the *time*.

> Social democracy works on the assumption that time is on our side. There must be plenty of it. Then one can move slowly towards the good society, step after incremental step, without having to clash head-on with the class enemy and break up its power; it will rather leak away in drips. But if catastrophe strikes, and if it is the status quo that produces it, then the reformist calendar is shredded. Social democracy can now do one of two things. It can continue to flow with the time, deeper into catastrophe – the choice from August 1914 – or it can become something else, another taxon of socialism, one that recognises that time is up and another decade or even year of this status quo is intolerable.
>
> *(Malm 2020:75)*

Today, social democracy is a politically ambivalent project, its right wing largely neoliberalized, its left wing clinging to the hope for incremental reform yet open to a socialist alternative. That left wing has served as a key protagonist for Green New Deals.

Green New Deal and Just Transition

In the United States, the Green New Deal (GND) burst onto the political scene in 2018, when organizers held a sit-in in the office of House Speaker

Nancy Pelosi. Although Pelosi quickly dismissed the idea as a 'green dream', since that time the GND has been at the centre of reformist climate action (Klein 2019). Like the New Deal of the 1930s, the GND is a social-democratic project, emphasizing the need to reduce advanced capitalism's massive social inequities, but unlike the earlier project, the GND directly addresses the ecological crisis. In a GND, states play an active role in steering investment away from fossils, and in providing a 'just transition' to displaced workers and communities affected by socio-ecological transformation. Although initially formulated as a national project, as Robert Pollin remarks, the scale of the climate crisis obviously requires coordinated climate action. He advances a global GND consisting of four components:

> 1 Phasing out global fossil fuel consumption by 2050;
> 2 Clean energy investments, averaging about 2.5 percent of global GDP per year, including both public and private investments;
> 3 Just transition support for workers and communities that are currently dependent on the fossil fuel industry; and
> 4 Phasing out deforestation and industrial agriculture, to be replaced with afforestation and sustainable agricultural practices.
>
> *(2023:143)*

Pollin envisages national economies (particularly in the global South) continuing to grow by replacing fossil fuels with clean energy. The latter (much of it devoted to energy efficiencies) would be developed in equal measure by capitalist and state investment.

Pollin recognizes that 'building a global clean energy infrastructure will entail a massive expansion in demand for the set of minerals that are used intensively in clean energy technologies;' and although the location of most of those minerals in the global South poses issues of further environmental degradation and imperialism, he is confident that supply issues can be mitigated by recycling minerals (2023:151). The other key aspect of a global GND is a just transition: a clean energy transition can create good, green jobs. Although there is no guarantee that such jobs will provide decent compensation and although countries with large fossil-fuel sectors will face greater challenges as jobs in carbon extraction disappear, overall, large investments in new green infrastructure 'will create increased leverage for political mobilization across the board—for improving job quality, expanded union coverage and more jobs for underrepresented groups' (2023:155). Key to a just transition is to protect energy-sector workers against major losses in living standards, through guarantees of new jobs at comparable wages and with intact pension provisions, with support extending to areas of retraining, job search and relocation. Finally, to finance the $2.8 trillion in yearly investment

Pollin estimates is necessary to reach net zero emissions by mid-century, a global GND would erect a global financing framework around carbon taxes (and rebates to ensure equity), downsizing military investment, and green bonds issued by central banks. Pollin's pragmatic GND 'will not replace capitalism with socialism. In fact, this variant of a global Green New Deal project will actually need to take root and flourish within the interstices of capitalism' (2023:163), but it will shift priorities from capitalist acquisitiveness to ecological sanity and egalitarianism.

How does this project address the trifecta of power? Regarding accumulation, as it operates in capitalism's interstices, the GND offers no frontal challenge to the structural power of capital, although increased public investment, including financing, alongside job creation, could shift the balance of power marginally. 'Green bonds' issued by Northern central banks but distributed to projects in the global South would stimulate green growth there; however, a global GND would not directly challenge eco-imperialism. Actually, resort to powering an ever-expanding global capitalism with renewables would inevitably increase extractive accumulation. The hegemony of capital would not be challenged either. A GND, if successful, could be expected to strengthen public support for progressive policy, weakening neoliberal common sense, but it would not address the hegemony of growth-as-accumulation, nor would it challenge other elements of capitalist hegemony rooted in the atomizing impact of ubiquitous markets and the persuasive power of corporate media.

In substance, GND proposals are social-democratic renditions of Climate Capitalism (Asher 2020), blending Keynesian economic management with market-based ecological modernization. Reviving a politics of class compromise as an alternative to neoliberal austerity, these reforms retain a faith in capital accumulation as the motor of progress, with change achieved through liberal-democratic measures (Nenning et al. 2023:5), rather than through new participatory sites of democratic planning. As Steve Fraser concludes, a GND is better than no deal, 'but it also assumes the limitless accumulation of capital on into a future not fundamentally different than what came before' (2024). As a result of GND reforms, the imperial mode of living may acquire a greener sheen, but since green Keynesian solutions are premised on uneven capitalist development and consolidated state power (Asher 2020:445), eco-imperialism will remain in place.

As with any political project, the devil is in the details. For Green New Deals, a basic issue is how a 'just transition' can be brought about, and what its scope will be. The standard framing of just transition 'simply entails a more interventionist state to create "green" jobs, internalizing the costs of capitalism's "negative social externalities" and providing a more adequate welfare safety net' (Albert 2021:93). Simone Abram and her coauthors

(2022) emphasize the need for a 'whole systems approach' to just transition, avoiding the reductionist and managerial language in which it is often framed, and taking up the multiple dimensions of justice – distributive, procedural, recognitional and restorative – that point towards transformation of the labour-capital relation itself. Ultimately, as the Labour Committee of Toronto-based Socialist Project has stated (Socialist Project 2008), 'The choice is not between jobs or the environment, or even the creation of a distinct "green" sector of the economy, but rather a transformation of all production and consumption so that we can satisfy our needs in environmentally sustainable ways.' In South Africa, the Right to Say No movement is pressing for a just transition that includes a new law guaranteeing communities 'the right to oppose business activities that negatively impact their lives and environment' (Pier and Hlabane 2024:121). Within the strategically crucial fossil-fuel sector, such a transition would likely mean fossil fuel nationalization to create 'an ownership structure in which it is possible to plan a rapid phase-out of fossil fuels and to transition fossil fuel workers into careers in renewable energy and related sectors' (Gunderson and Fyock 2022:383). The Just Transition is thus an object of hegemonic struggle over its actual form and content. Capitalist interests will respond not only with attempts to narrow and slow any transformation that may threaten profits; they will seek 'to "remake" just transition in their image through appeals to the 'common good' – a favourite trope in hegemonic struggle (Goods 2022:2129). This aspect of passive revolution has been on display in the German coal phase-out, which built on Germany's corporatist structure in an 'asymmetrical compromise' that prioritized capitalist accumulation over achieving democratic legitimation, and demoted the environmental element to a subordinate position (Haas et al. 2022:394).

Degrowth

Degrowth ideas in the global North have been in circulation at least since the 1970s, when ground-breaking works such as the Club of Rome's *The Limits to Growth* (Meadows et al. 1972) and Herman Daly's *Toward a Steady-State Economy* (1973) and *Steady-State Economics* (1977) appeared. Daly, a well-read economic historian and ardent critic of mainstream economics, introduced the idea of the embedded economy, 'an open subsystem of a finite and non-growing planet that was constantly supplied with low entropy energy from the sun and had to dissipate high entropy heat waste' – with global warming a consequence of failing to do so (Klitgaard 2023:94). Based on these foundations of environmental political economy, degrowth literature has blossomed in the 21st century. Largely the product of professional ecologists, this literature has unfolded with a faith in the power of a strong argument within liberal-democratic politics. Degrowth

> contains little on the resistance the transformation will encounter, not only from workers and consumers, who see their provisions reduced, but from the power of the capitalist class to resist any limitations on their power to accumulate. We should expect such a pushback from capitalists themselves, from a barrage of advertising and media, and from hired politicians.
>
> *(Klitgaard 2023:97)*

The most important insight in the literature on degrowth is the urgent need for an energy descent, indeed, for a decrease in the total material throughput of global production. In this, degrowth closely follows the critique of 'green growth' and Climate Capitalism I reviewed in the previous chapter. Hubert Buch-Hansen and Martin Carstensen (2021:322) outline four important contrasts between green growth and degrowth:

> The green growth project envisions changes within the framework of capitalism, it regards market actors to be pivotal for sustainability transitions, it is grounded in mainstream (environmental) economics, and while it works to the benefit of 'green' capital, its distributional consequences are overall modest. Conversely, the degrowth project envisions systemic change to bring about a profoundly different social order, it necessitates a citizen-led transformation, it finds scientific legitimacy in the contender field of social ecological economics and it would entail a far-reaching, global redistribution of economic resources.

As Stan Cox (2023) puts things, 'the only way that we humans can live within nature's resource restraints and ecological boundaries is to redirect our economies toward meeting all people's basic needs, and away from producing material overabundance. We have no choice but to converge on an equitable, modest level of energy and resource use that's enough to provide a decent life for all.'

Jason Hickel, a key proponent of degrowth who has served as an advisor to the Green New Deal for Europe, features energy descent in his definition of degrowth:

> Capitalism is a giant energy-sucking machine. In order to reduce energy use, we need to slow it all down. Slow down the mad pace of extraction, production and waste, and slow down the mad pace of our lives.
>
> This is what we mean by 'degrowth'. Again, degrowth is *not* about reducing GDP. It is about reducing the material and energy throughput of the economy to bring it back into balance with the living world, while distributing income and resources more fairly, liberating people from needless work, and investing in the public goods that people need to thrive. It is the first step toward a more ecological civilisation.
>
> *(2020:206)*

Hickel emphasizes the impossibility of dematerializing production and decoupling energy throughput from fossil sources. The empirical trends are clear. From 1990 through at least 2017 the correlation between global GDP growth and growth in global material throughput was nearly perfect. The Great Acceleration after World War 2 bears careful scrutiny. As the ideology of growth became entrenched, in tandem with accelerating economic expansion, material throughput exploded, reaching 35 billion tonnes by 1980 and 92 billion by 2017 – this when scientists estimate that Earth can cope with a total material throughput of approximately 50 billion tonnes per year. Nearly all of the ecological overshoot is driven by the imperial mode of living, that is, what Hickel terms 'excess consumption in high-income nations – consumption that is organised not around use-value but exchange-value' (2020:102). Strikingly, since 1980 the lion's share of new income from global economic expansion has gone to the world's rich.

As part of his critique of the green growth 'fantasy', Hickel criticizes the 'circular economy', which imagines that recycling can become a universal, eliminating the need for (most) extractivism. The aspiration to a more circular economy is of course welcome. But the idea that recycling will save capitalism is fallacious, since most material throughput cannot be recycled: 'in the end, only a small fraction of our total material use has circular potential' (2020:158).

As a radical vision, degrowth goes further than Green New Deals in its insistence on rejecting both the capitalist growth imperative and the sanctity of the commodity. Hickel asserts that

> by decommodifying public goods, expanding the commons, shortening the working week and reducing inequality, we can enable people to access the goods that they need to live well without requiring additional growth in order to do so. People would be able to work less without any loss to their well-being, thus producing less unnecessary stuff and generating less pressure for unnecessary consumption elsewhere.
>
> *(2020:235)*

Hickel's specific proposals for degrowth include an end to planned obsolescence; cuts to corporate advertising; a shift from individual commodity ownership to shared usership, an end to food waste; and a scaling down of ecologically destructive industries, along with a reduced workweek, decommodification of basic goods and a wealth tax (2020:209–31). These proposals are worthwhile, yet since they would retain capitalist control of much of economic life, it is unclear that they would create a radical break with capital's growth imperative. As Ralph Callebert points out, the typical argument for a reduced workweek reassures investors that no reduction in labour productivity (and thus profitability) would be risked. Such a reform could slow growth, 'in a more restrained and regulated form of capitalism', but would not mark a serious shift to degrowth (2023:138). Degrowth thinkers have been ambivalent on the issue of whether capitalism itself is compatible with degrowth. Advocates like Tim Jackson (2009) have suggested that reduced workweeks and the like 'could allow for a non-growing form of capitalism' (Klitgaard 2023:96).

Jason Hickel, however, takes a more eco-socialist, and anti-imperialist, position, demonstrating that the left wing of degrowth opens onto an eco-socialist project, to be discussed in Chapter 6. His critique of GDP as a reified metric of aggregate capital accumulation is paired with an understanding of innovation that cleaves it from the ideology of growth.

> We don't need *aggregate* growth to deliver innovation. If the objective is to achieve specific kinds of innovation, then it makes more sense to invest in those directly, or incentivise investment with targeted policy measures, rather than grow the whole economy indiscriminately and hope it will deliver the innovation we want.
>
> *(2020:200)*

More recently, Hickel has gone further in both his critique of capital and his direct advocacy of eco-socialism, with a formulation that moves closer to a historical materialist conception of green forces of production, as discussed in Chapter 1:

> We must be clear about what growth actually is. It is *not* innovation, or social progress, or improvements in well-being. It is very narrowly defined as an increase in aggregate production, as measured in *market prices* (GDP). GDP makes no distinction between $100 worth of tear gas and $100 worth of health care. This metric is not intended to measure what is important for people, but rather what is important for capitalism. Of course, what is important for capitalism is not to meet human needs, or achieve social progress, but rather to maximize and accumulate capital. If social progress and well-being are our goals, it is not the *market value of aggregate production* that matters but rather *what* we are producing (tear gas or health care?), and whether people have *access* to essential goods and services (is the health care privatized or universal?). This is basic to socialist thought. ... It should be clear from the above that degrowth is best understood as an element *within* a broader struggle for ecosocialist (and anti-imperialist) transformation. We must achieve democratic control over finance, production, and innovation, as well as organize it around both social and ecological objectives.
>
> *(Hickel 2023:46, 48, emphasis in original)*

Competing within degrowth politics with this socialist vision is a strong anarchist current (Trainer 2023). Oscar Berglund and David Bailey suggest that degrowth's roots in the 'newer social movements', compared to GND's base in unions and social democracy, lend the former a 'radical', anti-statist stance, discernable in its intentional regenerative communities, workers' cooperatives, community gardens and 'collective squatted housing projects.' Informed by the principles of prefigurative politics and horizontalism, each community project endeavours 'to instantiate future visions of society in small-scale initiatives in the present, or what are sometimes referred to as "nowtopias"' (2023:1019). The principle underlying these projects is that through ecologically sound and cooperative practices group members produce 'materially sustainable and beneficial outcomes', in cheaper food or energy (2023:1022). Some initiatives seek to scale-up their sustainable production by integrating it within existing processes; others strive 'to exit mainstream society altogether, often through the creation of an "ecovillage", in order to create new societies built on different and more sustainable foundations' 2023 (ibid.). The conclusion these authors draw is instructive. The debate between the GND and degrowth

> highlights some of the key conflicting assumptions held by social democrats/socialists, on one hand, who tend to see a central role for the state, and those adopting a more radical stance towards capitalist social relations, on the other hand, who are far more sceptical regarding the potential for state-based political activity to be able to deliver its promise of 'green growth.'
>
> *(2023:1027)*

Recalling Wainwright and Mann's contrasting ideal types, discussed at the beginning of this chapter, in this appraisal the binary between 'social democrats/socialists' and 'radicals' creates a Hobson's choice. Either climate activism can embrace statist reform that ignores capitalism's relations of production and thus allows endless growth, or it can embrace a radical project of prefigurative politics, creating new, sustainable relations of production at small scale, and ignoring the state. Conflating socialism with social democracy and reifying the capitalist state as impermeable to transformation, this framing imposes an unhelpful dilemma between co-optation and retreatism. Alex Callinicos has cautioned against the latter, anarchist option, which

> forgets that, if we ignore the state, it doesn't follow that the state will ignore us. Resistance as exodus carries the promise that we can cultivate our own garden, that we can find a space where we can live despite capitalism. But capital today, vigorously aided by the state, is invading the gardens of the world and sowing them with genetically modified crops.
>
> *(2006:256)*

A Gramscian emphasis on 'walking on both legs' (Carroll 2006:33) – contesting state power while creating change in civil society and in the mode of production – offers an escape hatch from this cul-de-sac. In the next chapter, I will elaborate an eco-socialist option, pointing to the need for a left party capable of unifying the many colours of progressive politics and *transforming/transcending* rather than *inheriting* the capitalist state.

Clearly, degrowth politics is not unitary. Dorothea Schoppek's analysis of the two discursive strands she discerns within it addresses whether the degrowth movement 'is more likely to become a counter-hegemonic project or serve as a passive revolution within neoliberalism' (2020:133). The 'sufficiency-oriented' strand calls for reduced consumption and increased self-production and sharing, emphasizing individual responsibility as both cause of and 'solution' to the ecological crisis. This framing resonates with the neoliberal project of 'responsibilizing' the individual, rendering degrowth vulnerable to co-optation within passive revolution. The 'practical left' strand combines the vision of a solidarity-based economy with re-embedding the economy into society and opening up a degree of freedom to act differently. Compared to the sufficiency-oriented strand, this strand foregrounds an ethic of solidarity (not simply sufficiency) and connects with a structural critique of capital, which can inform strategic action. The combination of structural and ethical critique calls for reflection and self-empowerment in order to create change, emphasizing micro-level direct actions 'to subvert the hegemonic formation by transforming its subjects' (2020:145). Schoppek argues that to achieve structural effects micro-struggles must be scaled up and combined with those on a macro-level – thereby avoiding relegation to a niche existence, as in cultivating 'our own garden.' For Schoppek, 'this means that everyday resistance struggles eventually have to be brought together in a concerted political project,' that 'radical transformation requires collective actions and an extroverted form of politicisation alongside self-transformations' (2020:146).

Buen Vivir

The alternatives to Climate Capitalism I have considered so far all have their origins in the movements and scholarship of the global North. An inspiring alternative, emanating from the South, which challenges the imperialist dimension of Fossil Capitalism from an Indigenous standpoint, goes by the moniker *Buen Vivir*, a Spanish translation of the Quechan concept *sumak kawsay*, which means living well. Embedded within a cultural matrix emphasizing the relationality of humans with each other and with nonhuman nature, the term has its cousins in Africa (*Ubuntu*, meaning a sense of communal reciprocity), in India (*swaraj*, meaning radical ecological democracy) and elsewhere (Acosta 2020:90–1). Although its roots are deep in Andean culture, *sumak kawsay* was systematized in Ecuador in the 1990s by the Kichwa Amazonian nation 'as a core principle for their territorial organisation in the context of struggling against the oil industry' (Malo Larrea et al. 2024:2).

Alberto Acosta served as Ecuador's minister of energy and mining and chaired the constituent assembly that established protection for the rights of nature in its constitution (effective in 2020). He suggests that 'we can understand Buen Vivir to be persons living in harmony with themselves, with other people in the community, harmony within the community and between humans and nature' (2020:89). This worldview differs fundamentally from the project of capitalist development. Emanating from 'indigenous people's modes of living', 'it is not part of an expansive and accumulative logic of progress and infinite growth, but rather seeks plenitude in balance and sufficiency' (Lang 2022:1287). As Miriam Lang elaborates,

> the communitarian modes of living anchored in sumak kawsay are *dysfunctional* to the capitalist logics of accumulation.... Instead of competition, sumak kawsay proposes collaboration. Instead of the capitalist *homo oeconomicus*, always rationally interested in getting the best out of everything for just himself..., it proposes an ontology of being collectively, in community, in awareness of our deep interdependences with others and our surroundings. Sumak kawsay understands that life is only possible on the basis of this web of relations.
>
> *(2022:1291)*

These values accord with degrowth ideas; indeed, the latter can be read as attempting to decolonize hegemonic narratives of progress 'by valorizing alternative conceptions of the good life' (Albert 2021:96). However, in Buen Vivir, the understanding of living well extends to a rejection of capitalist relations of production and an insistence on rebuilding solidaristic human community, which aligns it decidedly with the practical-left strand within Degrowth, and with historical materialism.[1] The capacious conception of living well is conveyed in three criterial dimensions that Marco Castillo has delineated, namely, interconnectedness, human well-being and harmony with nature. Buen Vivir 'connotes the importance of a fundamental respect for and harmony with nature, reflects a worldview where community, rather than individuality, is the focus of attention, and understands the economy as one based on solidarity where reciprocity takes precedence over the accumulation of wealth' (Castillo 2022:342).

In Ecuador, the movement dared to challenge neoliberal hegemony, 'seeking to forge a new future based on an organic vision of what the goals of human life and, by extension, the goals of government should be' (Castillo 2022:357). However, during the Buen Vivir government in Ecuador, 'the extractive economy remained and became more dominant' instead of shifting towards an intrinsic valuing of nature (Boogaard and Norren 2021:111). This outcome is understandable when one considers the hybrid character of Buen Vivir, as it developed since the turn of the 21st century through the efforts of organic intellectuals 'close to the grassroots' in Ecuador and Bolivia (Beling et al. 2021:25). A critical discourse analysis covering this period reveals three strands within Buen Vivir: the Indigenist strand (focused on avoiding exogenous, 'colonial' interference in its own territories), a state-centred, reformist strand, and a socialist strand – the last 'an attempt to assimilate globally circulating (neo)Marxist and neo-Keynesian discourses' into Buen Vivir (Beling et al. 2021:27). In Ecuador, statist reforms around Buen Vivir were increasingly framed in a conciliatory fashion vis-à-vis mainstream development strategies, as growing petroleum exports were 'justified with the argument that "more extractivism is needed to finance the transition out of extractivism"' (Beling et al. 2021:26).

Similar to Green New Deal and Degrowth, Buen Vivir is not a homogeneous project. In fact,

> buen vivir is neither a neo-ethnodevelopmental discourse pouring indigenous worldviews into the global public sphere nor a lineal one analogous to any quantifiable Western conception of well-being that can be seamlessly assimilated into existing bureaucratic structures and rationalities. Rather, its genealogical reconstruction as a spatiotemporally situated discourse has shown that buen vivir is rather to be understood as a "glocal" field of contention whose (limited) discursive variations can be traced to concrete agent-constellations and struggles in a context of global and local contestation around the prevailing model of development.
>
> *(Beling et al. 2021:28)*

Although rooted in traditional Indigenous cosmologies, Buen Vivir should not be reified. Through political co-construction involving local and global elements, it has forked into distinct currents. Going forward, the key challenge lies in identifying conditions under which Buen Vivir can generate 'broader convergences' towards 'transformative pathways' (or, at least, reformist approaches) to advance 'the transition to a fairer and more sustainable world' (Beling et al. 2021:29). As Adrián Beling and his colleagues conclude, 'the main value-added of buen vivir comes neither from its "retrotopian" significations nor from its (in)efficacy as a government program but from its politically and culturally subversive character, which produces an epistemic break with dominant languages and mind-frames' (2021:28).

The question for Buen Vivir is similar to the question posed by socialist revolutionaries regarding national liberation movements of the 20th century (Bowring 2021; Carlson 2023). Like contemporary initiatives informed by Buen Vivir, those anti-imperialist struggles fought for autonomy, for self-determination for colonized peoples. Once successful in the basic objective, national liberation was faced with the question of whether to remain within global capitalism or to push further, to a transformation of the actual mode of living. Most countries, under the thumb of British and US imperialism, were steered towards neocolonialism, but some (China and Cuba, for instance) continued the struggle for revolutionary transformation. Reflecting on the strands that comprise Buen Vivir today, the state-centred, reformist strand points towards something resembling a Green New Deal while the Indiginist strand points towards decolonization by de-linking. The socialist strand points towards articulating Buen Vivir with the wider, global struggle for a world beyond Fossil Capitalism. Such articulation should not be confused with dissolution of Indigeneity. What is needed is a process of mutual learning in solidarity, so that Buen Vivir and anti-colonial struggle are understood and embraced by non-Indigenous people as core to the left, while socialism and anti-capitalist struggle is understood and embraced by Indigenous people as core to Buen Vivir. In contrast, the Indiginist strand's quest to reconstruct autonomous communities implies local, decommodifying practices resembling those of Degrowth retreatism. Whether as Degrowth eco-villages or as resurgent Indigenous communities practicing Buen Vivir, such prefigurative formations do not challenge capitalism directly but strive to develop within its interstices. They could, however, function discursively as inspiration to a life beyond capital.

Like the Green New Deal and Degrowth, Buen Vivir includes strands of counter-hegemonic politics as well as currents that can be contained fairly readily within the passive revolution of Climate Capitalism. In our current setting, the challenge is to braid the counter-hegemonic stands into a powerful, transformative project informed by historical-materialist insights. The project that can actually move us to a life beyond capital, eco-socialism, is the focus of our final chapter.

Note

1 As Lebowitz observes, 'Capital, in short, constantly drives to crowd out all traces of the system of community' (2020:122), replacing solidarity with atomizing market transactions. The struggle against capital is a struggle for community.

References

Abram, Simone, Ed Atkins, Alix Dietzel, Kirsten Jenkins, Lorna Kiamba, Joshua Kirshner, Julia Kreienkamp, Karen Parkhill, Tom Pegram, and Lara M. Santos Ayllón. 2022. "Just Transition: A Whole-Systems Approach to Decarbonisation." *Climate Policy* 22(8):1033–49. doi: 10.1080/14693062.2022.2108365

Acosta, Alberto. 2020. "Buen Vivir: A Perspective for Rethinking the World." Pp. 87–99 in *Degrowth in Movements: Exploring Pathways for Transformation*, edited by C. Burkhart, M. Schmelzer, and N. Treu. Winchester, UK: Zero Books.

Albert, Michael J. 2021. "The Climate Crisis, Renewable Energy, and the Changing Landscape of Global Energy Politics." *Alternatives: Global, Local, Political* 46(3):89–98. doi: 10.1177/03043754211040698

Asher, Kiran. 2020. "The Brilliant, Monochromatic Red of *Climate Leviathan*." *Rethinking Marxism* 32(4):442–50. doi: 10.1080/08935696.2020.1807236

Azmanova, Albena. 2019. "The Paradox of Emancipation: Populism, Democracy and the Soul of the Left." *Philosophy & Social Criticism* 45(9–10):1186–1207. doi: 10.1177/0191453719872291

Beling, Adrián E., Ana Patricia Cubillo-Guevara, Julien Vanhulst, and Antonio Luis Hidalgo-Capitán. 2021. "*Buen Vivir* (Good Living): A 'Glocal' Genealogy of a Latin American Utopia for the World." *Latin American Perspectives* 48(3):17–34. doi: 10.1177/0094582X211009242

Berglund, Oscar, and David J. Bailey. 2023. "Whose System, What Change? A Critical Political Economy Approach to the UK Climate Movement." *Environmental Politics* 32(6):1012–32. doi: 10.1080/09644016.2022.2156179

Bernstein, Eduard. 1899. "Conclusion: Ultimate Aim and Tendency – Kant against Cant." *Evolutionary Socialism*. Retrieved March 7, 2024 (https://www.marxists.org/reference/archive/bernstein/works/1899/evsoc/ch04-conc.htm).

Boogaard, Birgit K., and Dorine E. van Norren. 2021. "'Development' Perspectives from the Global South: Learning from Ubuntu and Buen Vivir Philosophies." Pp. 104–16 in *The Politics of Knowledge in Inclusive Development and Innovation*, edited by D. Ludwig, B. K. Boogaard, P. Macnaghten, and C. Leeuwis. London: Routledge and Kegan Paul.

Bowring, Bill. 2021. "Marx, Engels, Lenin, and the Right of Peoples to Self-Determination in International Law." Pp. 98–114 in *Research Handbook on Law and Marxism*, edited by P. O'Connell and U. Özsu. Edward Elgar Publishing.

Bruff, Ian. 2014. "The Rise of Authoritarian Neoliberalism." *Rethinking Marxism* 26(1):113–29. doi: 10.1080/08935696.2013.843250

Buch-Hansen, Hubert, and Martin B. Carstensen. 2021. "Paradigms and the Political Economy of Ecopolitical Projects: Green Growth and Degrowth Compared." *Competition & Change* 25(3–4):308–27. doi: 10.1177/1024529420987528

Buci-Glucksmann, Christine. 1979. "State, Transition, and Passive Revolution." Pp. 207–36 in *Gramsci and Marxist Theory*, edited by C. Mouffe. Boston: Routledge and Kegan Paul.

Callebert, Ralph. 2023. "Changing Minds for a Changing Climate? On the Limits of How-To Projects for Saving the World." *Capitalism Nature Socialism* 34(2):134–41. doi: 10.1080/10455752.2023.2203529

Callinicos, Alex. 2006. *The Resources of Critique*. Cambridge, UK: Polity Press.

Carlson, John. 2023. *Empowering Indigenous Self-Determination In-Against-and-Beyond Capitalism: A Theory of Dual Dispossession*. Ottawa: Carleton University.

Carroll, William K. 2006. "Hegemony, Counter-Hegemony, Anti-Hegemony." *Socialist Studies/Études Socialistes* 2(2). doi: 10.18740/S44G7K

Carroll, William K. 2021. "Conclusion: Prospects for Energy Democracy in the Face of Passive Revolution." Pp. 479–504 in *Regime of Obstruction*, edited by W. K. Carroll. Edmonton: AU Press.

Castillo, M. A. 2022. "Implementing Indigenous Paradigms: The Paradoxes of Actualizing *Sumak Kawsay*." *Administrative Theory & Praxis* 44(4):340–62. doi: 10.1080/10841806.2022.2138196

Cox, Stan. 2023. "Even Plutocrats Can't Escape the Coming Heat." *Common Dreams*. Retrieved August 28, 2024 (https://www.commondreams.org/opinion/plutocrats-can-t-escape-climate-crisis).

Daly, Herman E. 1973. *Toward a Steady-State Economy*. San Francisco: WH Freeman & Co.

Daly, Herman E. 1977. *Steady-State Economics*. WH Freeman & Co.

Fraser, Steve. 2024. "The End of the Future." *Jacobin*. Retrieved 28 August, 2024 (https://jacobin.com/2024/03/left-politics-future-history-capitalism-progress).

Goods, Caleb. 2022. "How Business Challenges Climate Transformation: An Exploration of Just Transition and Industry Associations in Australia." *Review of International Political Economy* 29(6):2112–34. doi: 10.1080/09692290.2021.1956994

Gunderson, Ryan. 2023. "Powerless, Stupefied, and Repressed Actors Cannot Challenge Climate Change: Real Helplessness as a Barrier Between Environmental Concern and Action." *Journal for the Theory of Social Behaviour* 53(2):271–95. doi: 10.1111/jtsb.12366

Gunderson, Ryan, and Claiton Fyock. 2022. "Are Fossil Fuel CEOs Responsible for Climate Change? Social Structure and Criminal Law Approaches to Climate Litigation." *Journal of Environmental Studies and Sciences* 12(2):378–85. doi: 10.1007/s13412-021-00735-9

Haas, Tobias, Jeremias Herberg, and David Löw-Beer. 2022. "From Carbon Democracy to Post-Fossil Capitalism? The German Coal Phase-out as a Crossroads of Sustainability Politics." *Sustainability: Science, Practice and Policy* 18(1):384–99. doi: 10.1080/15487733.2022.2069542

Hansen, Mette Halskov, Hongtao Li, and Rune Svarverud. 2018. "Ecological Civilization: Interpreting the Chinese Past, Projecting the Global Future." *Global Environmental Change* 53:195–203. doi: 10.1016/j.gloenvcha.2018.09.014

Hickel, Jason. 2020. *Less Is More: How Degrowth Will Save the World*. New York: Random House.

Hickel, Jason. 2023. "On Technology and Degrowth." *Monthly Review* 44–50. doi: 10.14452/MR-075-03-2023-07_3

Huber, Matt. 2023. "The Left Is Losing the Climate Class War: Punishing the Workers Won't Save the Planet." *UnHerd*. Retrieved March 6, 2024 (https://unherd.com/2023/09/the-left-is-losing-the-climate-class-war/).

Jackson, Tim. 2009. *Prosperity Without Growth*. London: Earthscan.

Klein, Naomi. 2019. "Foreward." Pp. ix–xiii in *A Planet to Win: Why we need a Green New Deal*, edited by Kate Aronoff, Alyssa Battistoni and Daniel Aldana Cohen. London: Verso.

Klitgaard, Kent A. 2023. "Planning Degrowth: The Necessity, History, and Challenges." *Monthly Review* 85–98. doi: 10.14452/MR-075-03-2023-07_6

Lang, Miriam. 2022. "Buen Vivir as a Territorial Practice. Building a More Just and Sustainable Life through Interculturality." *Sustainability Science* 17(4):1287–99. doi: 10.1007/s11625-022-01130-1

Lebowitz, Michael A. 2020. *Between Capitalism and Community*. New York: Monthly Review Press.

Levrat, Virgile. 2020. "Sovereignty on a Changing Planet: Re-Reading Climate Leviathan." *Green European Journal*. Retrieved March 5, 2024 (https://www.greeneuropeanjournal.eu/sovereignty-on-a-changing-planet-re-reading-climate-leviathan/).

Lim, Audrea. 2018. "The Ideology of Fossil Fuels: Imagining a Low-Carbon World Means Revisiting Our Conception of Freedom Itself." *Dissent*. Retrieved February 7, 2024 (https://www.dissentmagazine.org/article/ideology-fossil-fuels-apocalypse-petrocapitalism-energy-humanities/).

Malm, Andreas. 2020. *Corona, Climate, Chronic Emergency: War Communism in the Twenty-First Century*. London: Verso.

Malo Larrea, A., M. Ambrosi De La Cadena, J. Collado Ruano, and L. Gallardo Fierro. 2024. "Transcending the Nature-Society Dichotomy: A Dialogue between the Sumak Kawsay and the Epistemology of Complexity." *Ecological Economics* 216:108044. doi: 10.1016/j.ecolecon.2023.108044

Meadows, D. H., Meadows, D.L., J. Randers, and W. W. Behrens III. 1972. *The Limits to Growth*. New York: Universe Books.

Nenning, Larissa, Paul Bridgen, Katharina Zimmermann, Milena Büchs, and Merita Mesiäislehto. 2023. "Climate Crisis and Social Protection – From Worker Protection to Post-Growth Transformation?" *Social Policy and Society* 22(4):695–714. doi: 10.1017/S1474746423000246

Norgaard, Kari M. 2011. *Living in Denial*. Cambridge, MA: MIT Press.

Parenti, Christian. 2022. "A Left Defense of Carbon Dioxide Removal: The State Must Be Forced to Deploy Civilization-Saving Technology." Pp. 130–42 in *Has it Come to This? The Promises and Perils of Geoengineering on the Brink*, edited by J.P. Sapinski, Holly Jean Buck, and Andreas Malm. New Brunswick, NJ: Rutgers University Press.

Pier, Lisa, and Matthews Hlabane. 2024. "Shifting Geographies of Extractivism in South Africa and Visions for a Just Transition from Below." in *State of Power 2024: Energy, Power and Transition*, edited by N. Buxton. Amsterdam: Transnational Institute.

Pollin, Robert. 2023. "The Political Economy of Saving the Planet." *The Japanese Political Economy* 49(2–3):141–68. doi: 10.1080/2329194X.2023.2262531

Schoppek, Dorothea Elena. 2020. "How Far Is Degrowth a Really Revolutionary Counter Movement to Neoliberalism?" *Environmental Values* 29(2):131–51. doi: 10.3197/096327119X15579936382491

Sen, Raka. 2019. "Book Review: Climate Leviathan: A Political Theory of Our Planetary Future." *Communication and the Public* 4(1):84–86. doi: 10.1177/2057047319836920

Simon, Roger. 1982. *Gramsci'a Political Thought*. London: Lawrence & Wishart.

Socialist Project. 2008. "Socialist Project Labour Movement Platform." Retrieved August 28, 2024 (https://socialistproject.ca/pamphlets/socialist-project-labour-movement-platform/).

Swyngedouw, Erik. 2013. "Apocalypse Now! Fear and Doomsday Pleasures." *Capitalism Nature Socialism* 24(1):9–18. doi: 10.1080/10455752.2012.759252

Taylor, Sarah McFarland. 2019. *Ecopiety: Green Media and the Dilemma of Environmental Virtue*. New York: New York University Press.

Teeple, Gary. 2008. *Globalization and the Decline of Social Reform: Into the Twenty-First Century*. 2nd ed. Aurora, Ont., Amherst, N.Y: Garamond Press, Humanity Books.

Tindall, David, Mark Stoddart, and Riley E. Dunlop, eds. 2022. *Handbook of Anti-Environmentalism*. Cheltenham, UK: Edward Elgar Publishing Ltd.

Trainer, Ted. 2023. "On Degrowth Strategy: The Simpler Way Perspective." *Environmental Values* 09632719231214309. doi: 10.1177/09632719231214309

Varoufakis, Yanis. 2024. *Democracy Is Finished in Europe*. Wave Media. Accessed August 28, 2024 (https://www.youtube.com/watch?v=2NYorC_8LXA).
Vaughan, Rachael A. 2021. "Why Bother? On Whether Individual Eco-Actions Count." *Ecopsychology* 13(4):276–83. doi: 10.1089/eco.2021.0001
Wainwright, Joel, and Geoff Mann. 2012. "Climate Leviathan." *Antipode* 45(1):1–22. doi: 10.1111/j.1467-8330.2012.01018.x
Wainwright, Joel, and Geoff Mann. 2018. *Climate Leviathan: A Political Theory of Our Planetary Future*. London: Verso.

6

TOWARDS ECO-SOCIALISM

Each of the projects I presented in the previous chapter offers some hope as we lean into what will be an epic struggle. Yet each, when viewed through the lens of historical materialism, misses key aspects of Fossil Capitalism's structure of domination. Let's take stock at this point, by posing three questions flowing from the discussion of historical materialism in Chapter 1. There, I introduced a trifecta of power, organized around the concepts accumulation, imperialism and hegemony:

- *Regarding accumulation*: In what ways does the project challenge the capitalist mode of production (both relations and forces of production), and present an alternative capable of addressing the climate emergency?
- *Regarding imperialism*: In what ways does the project challenge the geopolitical economy of ecological imperialism and the imperial mode of living premised upon it, and present an alternative that can provide a basis for global climate justice?
- *Regarding hegemony*: In what ways does the project challenge the hegemony of Fossil Capitalism and Climate Capitalism and provide an alternative hegemonic project capable of integrating a post-fossil historical bloc?

Green New Deals are effectively projects of Climate Capitalism from below. Operating within a social-democratic political logic, they seek to shift power and resources incrementally towards a regulated, post-Fossil Capitalism with a human face. Workers displaced in the restructuring would transition to jobs in growing sectors, and renewables would replace fossils in recomposing the forces of production. A global GND could generalize these practices, but the treadmill of production would continue to demand increasing resources

DOI: 10.4324/9781003412922-9

extracted mainly from the South. The GND is a project of the global North and has very little to offer the majority world.

In pursuing relatively modest objectives vis-a-vis climate breakdown, the GND faces two major challenges, the first relating to accumulation, the second to hegemony. Given capitalist control of economic resources (including the state's own tax base), 'fossil capital will be able to resist reforms to get off fossil fuels unless "despotic inroads" are made on private ownership and control of the means of production' (Chambers 2021:118). Otherwise, capital will go on strike, that is, withdraw investment, forcing political leadership to capitulate, or to begin a serious shift to socialism (Lebowitz 1995). As for hegemonic struggle, GND reforms replay the 'paradox of emancipation' to which social democracy has always been subject. Social democratic 'efforts at increasing equality and inclusion tend to validate and increase the desirability of the model within which equality and inclusion are being sought' (Azmanova 2019:1200). This was the paradox that trapped the Northern working class within the post-war class compromise. In some ways, GND politics exhibits 'a nostalgic longing for the pre-neo-liberal times of the Welfare State' (Azmanova 2019). To avoid its full incorporation into the passive revolution of Climate Capitalism, GND protagonists would need to address these challenges strategically. I agree with Vishwas Satgar that the issue on which GND politics could become transformative, and counter-hegemonic, is the 'just transition.' Most just transition schemes are shallow. They simply offer job re-training for displaced carbon-sector workers, replicating the clientelism of the welfare state, and the 'paradox of emancipation.' As an historical materialist, Satgar advocates a *deep* just transition. In his view, this means '1/ addressing the multiple, systemic crises of capitalist civilization, 2/ shifting from capital's growth principle to a principle of sustaining life, for the present and for future generations, and 3/ giving the transition a multilinearity, so that deep democratization occurs 'at different scales, locales and tempos, in workplaces, communities, civil society, on the internet (cyber democracy), and throughout the state and public sphere' (2018:64–5). Taking the idea of just transition seriously means rejecting Climate Capitalism and constructing a democratic way of life that is just and sustainable. This is a possible, eco-socialist trajectory for the Green New Deal.

In contrast to the GND, Degrowth directly addresses the capitalist growth imperative. But it does so in a confused way, and without much attention to capitalism's relations of production. Although climate mitigation requires lowering throughput and resource use, Degrowth focusses solely on the productive forces of capitalism without mentioning 'what the social relations of production underlying the alternative system will look at and how it will allow ... lower material abundance to be universally applicable and acceptable' (Chen 2023:166). The semantic equation *growth=accumulation=forces of production* leads to the mistaken conclusion that undoing capital's growth imperative requires diminishing the forces of production. As I discussed in Chapter 1, forces of production refer not to 'growth' or 'technology', but to the entire gamut of practices through which humanity purposefully engages with nonhuman nature. Graham (2023) has pursued the implications of this ecological understanding of Marx, noting (as did Marx) that science is a force of production – with the implication that ecological science and practice constitute an advance of these forces. An unappealing prospect for most people socialized into consumer-capitalist culture, 'degrowth' misconstrues what is actually needed, namely, 'a dialectical process of growth and degrowth in key sectors and productive forces based on democratic planning and public ownership' (Graham 2023:135). The mistake lies in conflating 'growth' with capital accumulation, leading Giorgos Kallis to conclude, 'no economic growth can be ecologically sustainable' (2017:190). Besides reproducing the nature/society binary, this way of thinking blurs the relationship between quantity and quality. Capital accumulation refers to the quantitative increase in abstract exchange value, and the metrics of capital – profit rates, GDP – reflect this; but qualitative advance in the material standard of living, and quality of life, is what matters to people. An eco-socialist approach, however,

> proposes *qualitative* distinctions. Some productions—for example, fossil energies, pesticides, nuclear submarines, and advertising—should not be merely reduced, but *suppressed.* Others, such as private cars, meat, and airplanes, should be *substantially reduced.* Still others, such as organic food, public means of transport, and carbon neutral housing, should be *developed.* The issue is not 'excessive consumption' in the abstract, but the prevalent mode of consumption, based as it is on conspicuous acquisition, massive waste, mercantile alienation, obsessive accumulation of goods, and the compulsive purchase of pseudo-novelties imposed by 'fashion.'
>
> *(Löwy (2023:156–7)*

For Güney Işıkara and Özgür Narin (2023:34), growth in the material standard of living includes 'a substantially higher amount of free time and the social and communal organization of reproductive labor, combined with the universal access to essential products in the broad sense.' Such gains, impossible under capitalism, are misplaced under the mantle of Degrowth.

Ultimately, as Jason Hickel recognizes, Degrowth (like the Green New Deal) is not a comprehensive project, but 'an element *within* a broader struggle for ecosocialist (and anti-imperialist) transformation' (2023:49). Originating in the professional class within global capitalism's core (Huber 2022), Degrowth is 'tailored for the Global North where an imperial mode of living prevails' (Huber 2022). Although its critique of extractivism calls attention to eco-imperialism, Degrowth 'is not a desirable and viable path for the Global South' (Huber 2022). It shares this Northern bias with the Green New Deal:

> Without a sufficient analysis of how colonialism and neocolonialism have impacted the Global South in the name of civilization, modernity, and development, the degrowth framework reveals the Eurocentric weakness that the Global Green New Deal also manifests. It is not enough to say Global South only needs to be supplied with technological and financial support in order for them to be incorporated into the Global Green New Deal project, just as it is not enough to claim that the degrowth project will excuse the Global South because they do not consume extra energy and resources. There needs to be a Southern-centered approach to a world-level degrowth project, and the South cannot once again play the role of a passive recipient of any consequences resulting from a North-initiated movement.
>
> *(Chen 2023:167)*

Finally, on the issue of hegemonic struggle, my view is that Degrowth is a very poor way of framing the alternative to Fossil (and Climate) Capitalism. As Stuart Hall (1988) noted years ago, the left must struggle for hegemony on terrain shaped primarily by the ruling historical bloc. The imperial mode of living – a reality for many in the global North and still an aspiration for many in the global South – forms part of that terrain, validating the virtues of consumer capitalism, and the indelible need for 'growth.'

In the circumstances, centring a counter-hegemonic project upon a vision of *subtraction* – what Huber (2022) calls a politics of less – is self-defeating. In a way similar to the Black Lives Matter call, in 2020, to 'defund the police', Degrowth projects a vision that appeals to movement insiders – what Smucker (2017:259) calls 'the righteous few' – but that creates obstacles to winning a great many hearts and minds. As I have argued above, 'degrowth' does not, in substance, even mean degrowth. It actually refers to development in sectors that enhance human and nonhuman welfare and the sunsetting of practices that are ultimately ecocidal. Fred Magdoff and John Bellamy Foster (2011) have termed this combination 'sustainable human development', a framing that resonates with the needs and aspirations of most of humanity.

This brings us to an important virtue in Buen Vivir. Its deep critique of capitalism coupled with the vision of a good life for all, within communities that care for people and nonhuman nature, does not begin with a rhetorical subtraction. 'Living well' is a compelling alternative to the frenetic compulsion to 'have more.' Emanating from the subaltern South, Buen Vivir delivers a strong critique of extractive imperialism, and the need to shift away from capital's growth imperative. Obviously, 'a good life for all' is a more resonant framing than 'degrowth', yet Buen Vivir's challenge to capitalist hegemony goes deeper, questioning the instrumentalization of nonhuman nature that has been central to the capitalist and colonial project. Without invoking Marxian vocabulary (and, it must be said, without a clear understanding of productive forces), Buen Vivir presents a critique of capitalism as a way of life.

As I mentioned earlier, Buen Vivir has forked into an Indigenist, state-centred reformist and socialist strands. If the state-reformist strand has tended to devolve towards Green New Deal politics, the Indigenist strand has a localist focus, integrally tied to the land and thus the lifeways (to be) reclaimed from the colonists. The goal is for each Indigenous community to create a permanent autonomous zone, detached from the maladies of global capitalism. But beyond the virtue of a good example, such 'postcapitalist localism' (Sharzer 2012:125–9) has little relevance in itself to the existential crisis humanity faces, and no decolonized Indigenous community will be protected from climate breakdown. In the struggle to prevent that catastrophe, Indigenous activism, reaching across national borders and working within coalitions (e.g., the Indigenous Environmental Network), is a significant force, and here the socialist strand of Buen Vivir shines. This strand proceeds from the strategic recognition that 'without mass movements for structural transformation of the colonial/capitalist state, Indigenous self-determination cannot be realised' (Dunbar-Ortiz 2016:86). Or, as Glen Coulthard (2013) has stated, 'for Indigenous nations to live, capitalism must die.' As I have concluded elsewhere, the left needs the communal vision and ethical sensibilities of Indigenous lifeways – encapsulated in Buen Vivir – but Indigenous peoples need historical materialist insights on capitalism and colonialism, 'as a resource for Indigenous resurgence and a counterweight to the lure of bourgeois modernization' (Carroll 2022:219).

Although Marx, Engels and some of their contemporaries[1] held strong ecological sensibilities, eco-socialism as a self-conscious perspective emerged in the late 20th century. My intent here is not to review what has become a profuse literature (cf. Baer 2018; Brownhill et al. 2022; Huber 2022; Saitō 2022), but to present eco-socialism as a perspective that can incorporate the best insights from the projects I have discussed in Chapter 5 into a viable counter-hegemonic strategy for socio-ecological transformation.

As an alternative to Fossil and Climate Capitalism based in historical materialism, eco-socialism is distinct in three respects. First, it views the dialectical relation between *forces and relations of production* as central to socio-ecological transformation. Second, it emphasizes the imperative to replace the anarchy of the market, capital's governing mechanism, with *democratic planning* at different scales, from local to global. Third, it identifies the social forces, already in motion but not politically integrated, that can be brought together to form an *historical bloc* capable of leading the transformation.

Transforming Forces and Relations of Production

Regarding the first issue, eco-socialism focusses on transforming both the forces and relations of production. As explained in Chapter 1, Fossil Capitalism is now a fetter upon the former, and in particular, upon the development and implementation of green forces of production needed to refuse ecocide (e.g., ago-ecology, ecological restoration; Graham 2023). Climate Capitalism, also expressing capitalist class interests but from a forward-looking rather than retrograde perspective, trumpets new technologies of renewable energy, CCUS and other forms of geo-engineering, to be rolled out through capitalist relations of production. Eco-socialism emphasizes the need for a double transformation, of productive forces and relations. It is not a matter of *shrinking* the forces of production but of *greening* them, and restoring a healthy metabolic relation between our species and nonhuman nature. Scientific innovation will be crucial in this. Holly Jean Buck's analysis of geoengineering is illuminating on this point:

> ...it is presumptuous to entirely cross off an idea that could, in a future scenario with runaway climate change, alleviate much suffering in places with less capacity to adapt to changing conditions. Systemic change is absolutely necessary. But geoengineering does not have to substitute for transformative change – in fact, to work well, geoengineering *requires* systemic change, because *responsible solar geoengineering requires carbon removal, which requires renewable energy*. This scale-up of renewables and carbon removal is only accomplishable with massive social and political transformation. The best-case solar geoengineering scenario is only achievable with dramatic social change. At the same time, critics rightfully worry that fossil fuel actors and other elites will use solar geoengineering to forestall social transformation, and the same concern applies to carbon removal. Yet at this point in time, a blanket rejection of carbon removal, in particular, comes off as an aesthetic luxury.
>
> *(2019:38–9)*

Given the accelerating pace of climate breakdown (and the climate system's enormous inertia, guaranteeing that CO_2 emissions remain in the atmosphere for a long time), effective climate-change mitigation will very likely require some combination of geoengineering initiatives, along with renewable energy and a reduction in material throughput. The greenhouse gases already released into the atmosphere will continue to heat the planet even as new emissions fall to zero. To bring greenhouse gas concentrations back to an ecologically healthy level, carbon dioxide will need to be removed from the atmosphere. The question is whether 'negative emissions' will serve the interests of capitalists (including fossil capitalists) in expanding their capital, or the interests of humanity, in a liveable world. Rolling out such new forces of production under capitalist control will simply cement another regime of accumulation, allowing another long wave of planet-wrecking exploitation and expropriation – which of course is the whole point in Climate Capitalism. For ethico-political *and for practical reasons*, these new productive forces must be introduced in the public interest, under democratic control. Technology in itself will not 'save the planet', but 'social and political practices might turn technologies into elements of long-term social projects for repairing, restoring, or renewing the biosphere' (Haines 2023:1344). For Christian Parenti, the need to adopt 'a radical approach to technology,' to become 'fully conscious environment makers', is urgent, and 'extreme technology under public ownership will be central to a socialist project of civilizational rescue' (2022:140). The 'disaster capitalism' – the further privatization of public resources in response to mounting environmental crises that Naomi Klein (2007) analyzed in the wake of Hurricane Katrina – must be countered with a 'disaster socialism'[2] that responds to the ecological crisis capitalism has created.

This means supplementing the core goal of a rapid fossil-fuel wind-down with remedial efforts to reduce atmospheric greenhouse gas. In moving beyond 'present capitalist perversions,' it also means 'refusing to let any negative emissions, imagined or real, ever justify continued "positive" ones' (Malm and Carton 2021:33). In Chapter 4, I pointed out that carbon-capture technology has not been scaled up to any meaningful level. But the current state of play should not be seen in static terms. For Andreas Malm and Wim Carton, the 'composite of circumstances makes it inadvisable to write off carbon dioxide removal completely' (2021:34). The prospects of direct air capture (DAC) as a means of capturing and storing carbon are real, even if the technology is not yet ready for large-scale application. As with renewable energy, which corporations are implementing even as fossil capital continues to accumulate, the crucial issue is who controls the technology, and to what ends. 'Renewable energy and DAC here inhabit the same technological battlespace. Fossil capital deploys them to reproduce itself; the task for any counterforces is to instead arrange them so as to maintain a habitable planet' (2021). Under capitalist control, DAC is poised to become a 'sewage system' for continued fossil-capital accumulation and a new means of profit-making in the business of 'air-mining' (2021:36). Malm and Carton agree with Buck 2021 (2019) and Parenti (2022) that the state is 'the sole actor with a potential to mobilise resources' for implementing DAC at scale (2021:36), a position I also find compelling.

This discussion of the need for green forces of production has already directed us to the relations of production, which strongly shape those forces. To address the current state of ecological overshoot, capitalist dominance within those relations must be countered with democratic, public control of our productive relations with each other. In our current circumstances, this means taking class politics seriously, as Matthew Huber has in *Climate Change as Class War*. A critic of Degrowth, Huber finds promise in a radicalized Green New Deal, including a deep just transition, to shift the balance of class forces while implementing climate mitigation policies. Emphasizing decommodification, a radicalized GND is about 'shifting power and control over society's resources. The most ecologically beneficial part of this program is that it aims to transfer key industries from private to public ownership so that environmental goals can predominate over profits' (2022:208). The comparison with Degrowth is helpful here. A radicalized GND, incorporating a class politics,

> would articulate a confrontational approach where the capitalist class must *degrow* so that the working class can see growth in material security and basic human freedom. The politics of degrowth at the aggregate in the name of ecology refuses this kind of antagonistic class politics where some lose but more gain.
>
> *(Huber 2022:169, emphasis added)*

An eco-socialist transformation of production relations proceeds, necessarily, from this shift in power and control, but goes beyond 'degrowing' the capitalist class and expanding space for the public interest. Ultimately, private, oligarchic control of the forces of production must give way to public and democratic control.

As Kohei Saitō has emphasized in his vision of 'degrowth communism', the enclosure of wealth, the creation of artificial scarcity central to commodification must give way to a 'commonification of wealth' (2022:229) – an abundance of common wealth produced and consumed in accordance with ecological wisdom. Inspired by Marx's *Critique of the Gotha Program* (2022 [1875]), where Marx outlined his own vision of post-capitalism, Saitō envisages a transformation in the relations of production, 'abolishing the excessive division of labour and making labour more democratic and attractive' in order to ensure 'the free and autonomous activity of individual workers' (2002:233). He points out that although such liberatory change could decrease labour productivity, the forces of production would actually expand. For historical materialism, the most important force of production is the human being; hence, the empowerment, rather than the diminishment, of workers within their labour processes is a fundamental gain in the forces of production. The key

> is the active participation of workers in deciding what, how and how much they produce. This democratic production is the direct antithesis of the 'despotic' character of capitalist production. Associated producers

> more actively participate in the decision-making process without the imposition of the will of the few. Hierarchal control is incompatible with Marx's vision of providing more autonomy to the associated producers, but without hierarchy, it takes more time to mediate between different opinions and reach a consensus.
>
> *(Saitō 2022:241)*

Time is an important consideration in all this. As Marx argued in 1847, with the subordination of worker to machine, within industrial capitalism's extreme division of labour, workers 'are effaced by their labor' (Marx 2009 [1847]:21) and reduced to simply another factor of production in the competitive quest for profit.

> Time is everything, man is nothing; he is, at the most, time's carcase. Quality no longer matters. Quantity alone decides everything; hour for hour, day for day...
>
> *(Marx 2022 [1875])*

In contrast, 'once the aim of social production is emancipated from the pressure of infinite capital accumulation, there is no need to produce an enormous and even wasteful amount of surplus products' (Saitō 2022:239). Reducing throughput (and extraction of non-renewable resources) and increasing time for self-development and community development expresses an enhancement of productive forces and a transformation of relations of production, within an ecologically responsible framework.[3]

Democratic Planning

Karl Marx wrote, in the first volume of *Capital*, that 'in the society where the capitalist mode of production prevails, anarchy in the social division of labour and despotism in the manufacturing division of labour mutually condition each other...' (1976:477). For Marx, capitalist relations of production include the market mediation of many capitals, as the capitalists controlling them compete for shares of surplus value. This is an anarchic process – unplanned and driven simply by the pursuit of profit by investors. At the same time, within each capitalist enterprise capital exercises despotic rule over segments of the working class, keeping the costs of production as low as possible by constraining wages, refusing reductions in the length of the working day, speeding-up production, introducing labour-saving technologies and externalizing whatever costs can be deflected. Capitalist relations of production consist of the dialectical combination of *despotism within firms* and *market-mediated anarchy among them*. Eco-socialism, in contrast, calls for democratic planning, both within production sites and in the wider economy.

We have seen that reliance on market mechanisms, as in Climate Capitalism, cannot induce radical changes in our socio-ecological relations within the tight timespan now left.[4] As John Bellamy Foster has put matters, in conversation with Batuhan Sarican,

> The myth of the self-regulating market system is exactly that, a myth. It becomes a justification for letting all outcomes arise ex post rather than ex ante, that is excluding all substantive planning, so that the capitalist class and corporations basically can mediate all developments and manipulate them to their own end under the guise of neutral 'market forces.' … It would be suicidal to leave the future of humanity up to so-called market forces, that is, global capital, which has only one aim: the endless accumulation of capital at the top of society, which has its counterpart in 'Après moi, le déluge!' Without planning controlled by the associated producers there is no way of preventing the runaway train of capitalism from taking us over the cliff. Planning, of course, does not mean the elimination of markets. It does mean that the economy would not be controlled by 'markets.' The truth is that the dominance of 'market forces' today simply means that monopoly-finance capital is left in charge: the very force that has both brought us to the brink of planetary ecological collapse and that is preventing us from doing anything about it.
>
> *(Foster and Sarican 2023)*

Extensive research on the power structure of contemporary corporate capitalism certainly shows that the main players, globally and at the national level, are giant transnational corporations and financial institutions. Controlling capitalists form a well-integrated elite, linked to each other by many extra-market social relations, from interlocking corporate directorates through think tanks and industry groups, to business councils and private clubs (Carroll 2010; Carroll et al. 2021; Carroll et al. 2023). Large corporations do compete with each other for market share (as in Coke vs. Pepsi), and for the good graces of investors (telegraphed in quarterly corporate results and share prices), but capitalism today is a far cry from a 'free market system.' The capitalist class's top tier actually comprises an oligarchy. However, capitalists themselves operate within a matrix of *systemic power*, which dictates to them a limited range of options consistent with the system-imposed objective of profit maximization. This systemic power of capital can be defined as '*capital's capacity to impose its logic on social life*; a capacity which includes and ultimately relies upon, yet is not reducible to, relations among social actors in a traditional sense, such as the relationship between capitalists and proletarians…' (Mau 2023:46, emphasis in original). In eco-socialism, planning replaces this systemic power.

Based on her analysis of 'cannibal capitalism', on which I drew earlier, Nancy Fraser offers a 'simple formula' for conceptualizing the roles of markets in a socialist society: 'no markets at the top, no markets at the bottom, but possibly some markets in the in-between' (2022:155–6). At the top, the social surplus, which shapes our collective future, 'must be allocated via collective processes of democratic planning in which market mechanisms should have no place.' This directly addresses capital's systemic power. At 'the bottom,' our basic subsistence (shelter, clothing, food, health care, education, transportation, communication, energy, leisure, clean water, and breathable air) should take the form of public goods. Their allocation must, again, 'be a subject for democratic discussion, contestation, and decision making – through which needs are met "as a matter of right, and not on the basis of ability to pay"' (2022). Fraser envisages 'the in-between' as 'a space for experimentation ... where markets could find a place, along with cooperatives, commons, self-organized associations, and self-managed projects' (2022). This vision places *public good* and *public provision* at the centre of a post-capitalist way of life, creating 'a different *logic* guiding production' on the basis of social need, including most urgently 'the need for a decarbonized energy system' (Huber 2022:291).

Planning should be seen as 'the negation of the anarchy of production' (Chen 2023:169). In ecological planning, 'society – based on scientific evidence and public debate – democratically decides how to organize the process of social provisioning and how to avoid transgressing planetary boundaries' (Schmelzer and Hofferberth 2023:143).[5] Eco-socialist planning extends the planning that already takes place within capitalist enterprises, while transforming it from a top-down expression of oligarchic power-over to a democratic form of power-with. Marx saw such an emergent capacity to consciously plan and coordinate production overall as a new productive force (Graham 2023:129). As we face looming climate catastrophe, two priorities stand out: 1/ the need to make planning as participatory as possible, avoiding the replication of top-down, bureaucratic practices and 2/ the need for 'globally coordinated planning' (Schmelzer and Hofferberth 2023:151) that moves beyond the legacy of colonialism and imperialism.

On the first point, any eco-socialist project faces the challenge of transforming the capitalist state it inherits into a socialist, democratic state. In Sam Gindin's socialist realist framework, this involves the conversion of state ministries into 'sectoral workers councils' comprised of delegates from the workplace collectives that replace capitalist firms, enabling workplace democracy. As much as possible, planning can be regionally devolved,[6] bringing the process closer to those affected by decisions, deepening community development and creating layers of democratic planning, from the workplace, through sectors and regions, to what Gindin calls the 'central planning board.' That democratically constituted board would control allocation of investment resources to sectoral councils and regions, within a framework of decentralized power that 'makes the importance of a coordinating body, even if less directly hands on, even more critical' (Gindin 2019:32). Gindin's socialist realism contrasts sharply with anarchist dreams of ending domination by abolishing the state *tout court*. 'If the state is seen as a set of specialized institutions that not only mediate social differences and oversee judicial discipline but also superintend the replacement of the hegemony of class and competitive markets with the democratic planning of the economy, then the state will likely play an even greater role under socialism' (2019:19).

Marta Harnecker and Michael Lebowitz, who advised the Chavez government in the early years of the Bolivarian revolution in Venezuela, offer similar insights on the practice of democratic planning. At the centre of democratic socialist planning is human development through popular participation, supplanting the atomization of life within late capitalism. 'Participation, protagonism in all spaces, is what will allow human beings to grow and increase their self-confidence, that is, facilitate human development' (Harnecker 2015:70). Yet,

> there must be a national strategic plan that coordinates local plans. Each of the decentralized spaces should be part of the national whole and be willing to contribute its own resources to strengthen the development of those spaces with the greatest shortages. This kind of decentralization must be imbued with a spirit of solidarity.
>
> *(2015:82)*

The practice of democratic planning needs to be approached dialectically, walking 'on two legs, in order to (a) take the old state away from capital; and (b) to build a new state through institutions such as workers' and communal councils that develop the capacities of the working class' (Lebowitz 2020:166).

Popular protagonism and subsidiarity, bringing decision-making to local scale where feasible, 'with a spirit of solidarity', should not be conflated with eco-localism. Popular among mainstream environmentalists, eco-localism 'projects the local as an ideal scale and conceives communitarian eco-utopias in a politics that is individualizing and particularizing' (Albo 2006:23). As the term 'personal fiefdom' implies, localism can be anything but democratic.

> Localness can only become an emancipatory virtue if it is nested in a broader, interwoven structure that is regulated and coordinated by collective bodies of workers. Interdependence of localities under socialism would not imply a power asymmetry or hierarchy between them, which is characteristic of capitalism, but rather represent the source of their collective power. For instance, anticipated disruptions in food production and projected shifts in agricultural practices due to the planetary crisis would not be met by local self-sufficiency under socialism, but rather, by a

> condensation of coordination at higher levels of planning, where a more global picture is available. The same can be said for the expansion and scaling up of agroecological food systems, ecological restoration, and earthcare labor.
>
> *(Işıkara and Narin 2023:38)*

Matt Huber (2022:287) insists that 'global production must be socially coordinated to stave off climate catastrophe.' There are good reasons for this verdict. The scale of the climate crisis obviously requires globally coordinated planning, to undo the ruinous impact of capitalist globalization. Climate change is the most urgent aspect of global ecological crisis. But the dramatic loss of biodiversity (the 6th Extinction we are now living through) and the degradation of local ecologies through mining, cattle ranching and other practices must also be addressed – along with the uneven geographical impact of the crisis, which is already creating food insecurity and climate refugees in regions suffering extreme degradation. Hence, the need for democratic planning of global resource flows and of globally coordinated ecological restoration (Candeias 2013). Again, eco-socialism in our era will be 'disaster socialism' – reversing the logic of 'disaster capitalism' that Naomi Klein exposed in *The Shock Doctrine* (2007). Disaster capitalism turns cataclysmic events that are in many cases symptoms of climate breakdown (e.g., superstorms, floods) into opportunities for accumulation by dispossession, via neoliberal privatization, including privatized disaster responses. Disaster socialism will respond to the same events by caring for and empowering the people victimized by such disasters and restoring damaged ecosystems. By the same token, any resort to geoengineering directed at the entire climate system must be undertaken through globally coordinated central planning, 'imbued with a spirit of solidarity' (Harnecker 2015:82).

Alongside eco-socialist disaster response, in addressing the climate crisis proactively, globally coordinated planning will need to follow the formula of 'contraction and convergence', the idea that 'every country must reduce its emissions and that all countries must converge on net zero emissions' (Maslin et al. 2023:2). In convergence, the advanced capitalist countries (responsible historically for most carbon pollution) reduce emissions at a much faster rate than global South countries (some of which may increase emissions in the short term). Such an approach is crucial in transcending the 'exploitative colonial and post-colonial relationships, processes, and institutional structures' of eco-imperialism, 'redistributing wealth and access to technologies, and making global and local consumption patterns more equitable' (Ciplet 2023:2) – moving us towards 'a good life for all within planetary boundaries' (O'Neill et al. 2018). As David Ciplet argues, effective and socially just climate governance will entail a blend of mitigation and adaptation to stabilize the climate system. The climate governance spectrum that Ciplet has outlined is a useful starting point (see Figure 6.1).

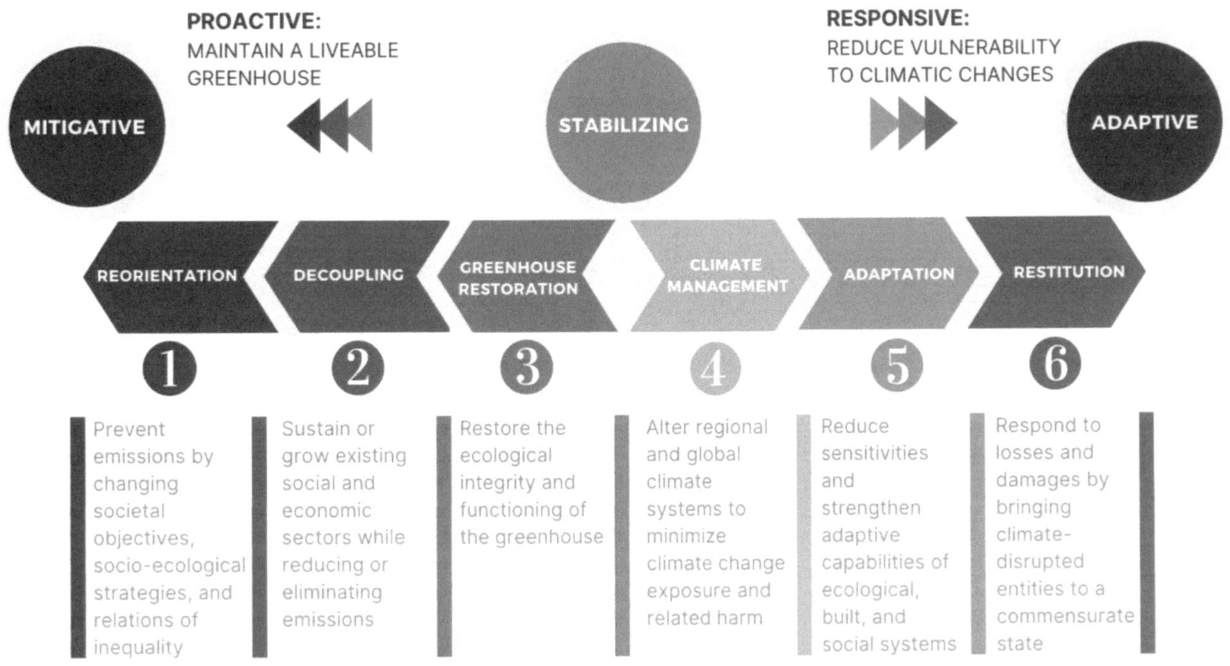

FIGURE 6.1 The climate governance spectrum and the six pillars of climate governance.

Source: Ciplet (2023).

Forming an Eco-socialist Historical Bloc

To transform the forces and relations of production, to shift from the anarchy of market-driven capitalism to a democratic, planned economy, requires collective agency. Who are the agents, what are the practices and organizational forms for this? These are questions for *hegemonic struggle*. In addressing them, I find the thought of Antonio Gramsci, the theorist of note on hegemony, particularly generative. In previous chapters I discussed the hegemonic struggles that gave rise, in the North, to carbon democracy in the early to mid-20th century (Chapter 1), to Fordist class compromise in the post-World War 2 decades (Chapter 2), and to post-Fordist neoliberal globalization in more recent decades (Chapter 3). In Chapter 4 I unpacked the passive-revolutionary agenda of Climate Capitalism, which strives to pivot from Fossil Capitalism to 'green growth' that preserves the economic nucleus of capitalist power. To oppose this passive revolution, we need an anti-passive revolution that actively advances eco-socialism as a hegemonic project resonant with broad sectors of global humanity, constituting an historical bloc capable of building eco-socialism.

In the conduct of such a hegemonic struggle, three issues stand out. First, there is the question of the *social forces* that can be brought into the bloc – those with an interest in ending Fossil Capitalism. Second, there is the question of which *practices* can most effectively produce transformative changes – in the current conjuncture and longer term – to disempower capital and capitalists while effectively addressing the ecological crisis. Third, there is the question of *organizational form*: how can the practices of those social forces – the collective protagonism opposing capitalism and seeking a just and liveable world – form an effective 'political instrument' for transformative change?

Social Forces for an Eco-socialist Transformation

On the first question, from an historical materialist perspective, the pivotal social force in the struggle for socialism is the proletariat.

> The power of the working class is rooted in three factors. First, it is the vast majority of the population – meaning any *democratic or majoritarian* approach to climate action must build a working-class coalition. Second, its strategic location at the point of production gives it structural power over the source of capital's profits and social reproduction more generally. Working-class power is most effective in periods of mass strikes and disruption that force elites and capitalists to cede to mass demands. Third, because economic insecurity defines working-class life, they have a fundamental material *interest* in transformations in the relations of production. *(Huber 2022:6–7)*

Yet, as John Bellamy Foster has observed in conversation with Jia Keqing, 'what is involved in class struggle today is not simply struggles in the workplace, though, as always, this is the center of working-class power, but also struggles over the whole environment. It is becoming more and more difficult to separate the economic and environmental conditions of material existence' (Foster and Keqing 2023:41). Foster, who introduced the concept of 'environmental proletariat' in 2010 (Foster 2010), goes on to remark that, particularly in the global South, economic and ecological problems

> are more and more intertwined given the structural crisis of capital and combined economic and ecological crisis and catastrophe. The economic proletariat has often been constrained by the logic of trade unions and the struggle for wages and benefits. The environmental proletariat, which is simply a way of referring to the proletariat in terms of the full complexity of its material existence, is concerned with work relations but also the full range of material life conditions. Such a unified standpoint is necessarily more revolutionary and more capable of grappling with the problems of the age. ... To speak of an environmental proletariat is thus to speak of a broader proletariat, the coming together of environmental and economic concerns, of proletarians, peasants, and the Indigenous. It means dealing with issues of social reproduction under capitalism that have led to extreme gender-based oppression of women.
>
> *(Foster and Keqing 2023:41)*

In the environmental proletariat, the politics of production and of socio-ecological reproduction are conjoined. As Fossil Capitalism has opened a widening metabolic rift, social forces opposing capital's domination of labour are aligning with forces opposing capital's domination of nonhuman nature. The close articulation of capitalist domination over labour with capitalist-driven degradation of the ecosystems that enable social reproduction is facilitating diverse alliances of movements. 'We are seeing, in short, the rise of a globalised environmental proletariat as a conscious class for itself, i.e., as a worker-community formation with a new ecological sociability, embracing a vision of human production in its most fundamental sense as the metabolism of nature and society' (Burkett 2020:93).

The political currents opposing Fossil Capitalism that I discussed in Chapter 5 – the Green New Deal/Just Transition, Degrowth, Buen Vivir – each contain elements of the environmental proletariat that are potentially inclined towards eco-socialism. The challenge lies in pulling these into a coherent project with a mass base. David Ciplet (2022) has addressed the conditions under which Just Transition coalitions can facilitate the formation of such a counter-hegemonic historical bloc. To transform political-economic structures, 'advocates must strategically contend with multi-dimensional nodes of power ... in civil society, political society, and economic society, and their relation to a just transition' (2022:320).

Noting the 'pitfalls of transition coalitions that are either impeded by their inability to gain strategic forms of power, or disembedded from the concerns and leadership of directly impacted communities,' he emphasizes the need to build

> transformative just transition coalitions ... in which diverse counter-hegemonic forces gain significant strategic power, while simultaneously embedding in the concerns, leadership, and decisionmaking of directly impacted peoples.
>
> *(2022:327)*

Indigenous communities, dispossessed from traditional lands by colonialism and imperialism, are an important part of the environmental proletariat. Buen Vivir, of course, is a highly embedded form of political agency, grounded in Indigenous lifeways and resistance to eco-imperialism, but to gain strategic power in civil, political and economic society the Indigenous current needs to flow together with other movements that share a vision of decolonization, decommodification and decarbonization – the 'three ds', as Eaton (2021) has characterized them. Degrowth, as we have seen, contains a 'sufficiency-oriented' strand, emphasizing individual responsibility, which resonates with neoliberal 'responsibilization', but also a 'practical left' strand that takes up an environmental-proletarian standpoint, advocating a re-embedded, solidarity-based economy (Schoppek, 2020). The former strand exemplifies what Schoppeck calls 'subhegemony', appearing to be counter-hegemonic, yet containing elements of hegemony that can have a 'flanking effect', thereby helping to solidify hegemonic power by deflecting attention from the need for structural transformation. For instance, in sufficiency-oriented degrowth movements, the emphasis on self-care, food-sharing and the like actually supports neoliberal hegemony by ignoring structural factors and atomizing individuals (Schoppek 2020). The same may be said of GND/Just Transition coalitions that are in tune with struggles on the ground (i.e., embedded) but lack the forms of strategic power, in the state, economy and civil society, necessary to translate their proposals into truly transformative change.

Clearly, there is a diverse array of social forces for whom a transition to eco-socialism would be enormously beneficial, even life-saving. The environmental proletariat comprises most of humanity, and aligned movements for climate and social justice can amplify its presence while extending the political agenda to issues of race, gender, sexuality, ability and other concerns that are integral to socialism. Yet the alignment is not yet organized in a politically impactful form – a unity-in-diversity. Moreover, as the example of subhegemonic flanking within Degrowth shows, neoliberal hegemony continues to persuade, often deflecting us from the steps we need to take collectively to avert ecocide, and towards practices that actually bolster the status quo.

Non-reformist Reforms and War of Position

This discussion takes us to the second question in building an alternative historical bloc: which *practices* can most effectively produce transformative changes – in the current conjuncture and longer term – to disempower capital and capitalists while effectively addressing the ecological crisis? Since publication of Naomi Klein's *This Changes Everything* the concept of 'Blockadia' has caught on among climate justice activists. Klein introduced the term to refer to 'a roving transnational conflict zone' of 'increasingly interconnected pockets of resistance' to 'high-risk extreme extraction', building a global movement 'driven by a desire for a deeper form of democracy' (2014:294–5). She pointed to the tendency for young people to eschew the top-down environmentalism of closed-door COP meetings, lobbying and the like, in favour of local resistance in the form of disruptive direct action, blocking extractive projects and protecting lands from becoming 'sacrifice zones' for big carbon. Klein argued that the 'friction' created by Blockadia and related forms of resistance is 'needed to put the brakes on the forces of destruction and destabilization' (2014:451). In British Columbia, where both Naomi Klein and I live, Blockadia has had a high profile. Protests, including blockades, against the Trans Mountain Pipeline Expansion (TMX) substantially delayed the project and spooked its American investors, leading the Canadian government to purchase the project in 2018. However, as I write this, TMX is filling with diluted bitumen (dilbit) and, within weeks, will begin transporting the fuel from a terminal on Canada's west coast. The state-owned pipeline will never turn a profit: its total construction costs ballooned from an estimated $7.4 billion in 2018 to $34 billion today (Bakx 2024). Instead, it stands as a massive subsidy to tar sands capital. Four hundred kilometres north of the TMX, on the unceded land of the Wet'suwet'en people, an Indigenous-led blockade has slowed construction of the BC Coastal Gaslink project, which will transport natural gas to one of six LNG facilities being constructed on the same west coast. The $40 billion project, the single largest private investment in Canadian history (CBC News 2023), is nearing completion, the pipeline having been entirely laid as of October 2023. After valiant and sustained direct-action resistance, these projects will soon add to the carbon pollution that is taking us towards ecocide. The same may be said of other Blockadia campaigns, such as the 2016–17 resistance to construction of the Dakota Access Pipeline, led by the Standing Rock Sioux, which however did not appreciably delay the project's completion. Each of these resistance campaigns created on-the-spot solidarities among activists, achieved extensive media coverage and met with militarized state violence that effectively neutralized opposition.

Resistance to carbon extraction, although an important aspect of climate-justice politics, cannot take us beyond Fossil Capitalism. As New Left veteran Tom Hayden stated, addressing anti-WTO activists at the 1999 Battle in Seattle,

> You have slowed the machinery of destruction down. But it can't be about slowing the rate of destruction. It has to be about speeding the rate of creation – of a new world, a better place.[7]

'Resistance is fertile' is a popular anarchist slogan, but new growth requires more than fertilization. Anarchism (including anarchist currents within Degrowth and Buen Vivir and direct-action groups like Extinction Rebellion) conceptualizes new growth as prefigurative politics – living differently in the here-and-now, by establishing local, non-hierarchical and communal practices in the interstices of late capitalism. However, this *retreatist* conception of prefiguration is narrow, and shallow.

> ...[E]xit is not enough. After we are out the door, then what?... If we expect others to take the leap to freedom with us we need a coherent and practical sense of what this freedom will look like on the other side of the door. Otherwise, we risk the fate of previous counter-cultures – well-meaning but marginal in their appeal. Easily isolated and swallowed up by the momentum of the system. Or brutally suppressed while uncomprehending majorities stand by and shake their heads.
>
> *(Swift 2016:149)*

If the repertoire of movement practices is restricted to direct-action resistance and local, grassroots prefiguration, the prospects of building a transformative historical bloc are quite slim. This approach reifies the state as a fixture impervious to transformation, whose in-built authoritarianism can only be resisted, while reducing prefiguration to what can be accomplished in the here-and-now. As Brecht De Smet points out, a self-contained, localist logic of prefiguration

> precludes a transfiguring moment. Because a prefiguration already constitutes the future society in a *developed* micro-form, it does not need to learn or to mature, but only to expand and be repeated. A prefiguration does not enter into a dialogue, but simply encourages non-participants to *imitate* and adopt the prefigured practice.
>
> *(2014:314)*

The climate crisis necessitates a 'transfiguring moment', and requires much more than proliferating and scaling up local projects. Although initiatives to decommodify lifeworlds have value in creating the subjective basis for wider change (what Gramsci termed intellectual and moral reformation), 'radical prefiguration needs to go beyond lifestyles and subcultures, to address system transformation, as in the creation of a solidarity economy' (Carroll 2015:664).

In the hegemonic struggle for hearts and minds, Fossil Capitalism (and Climate Capitalism) operate at different scales, securing consent through everyday practices (e.g., automobility), through local boosterism, within institutions such as universities and media, and through state organizations from the municipal to the global (see Table 4.1). In the war of position to disempower capital, efforts at socio-ecological transformation must be equally multi-scalar, and must aim not only at decolonization of lifeworlds (Habermas 1987) but at democratization of state and economy. As we have seen, incremental reforms typically feed into a paradox of emancipation: reforms give the system a more human face, which actually increases popular loyalty to the way of life it enables (Azmanova 2019). To avoid co-optation, counter-hegemonic practice needs to push for *non-reformist reforms*, and to link these reforms together in a multi-scalar and multi-frontal war of position.

Rather than functioning to maintain the system, non-reformist reforms 'create the conditions for deeper transformations' (Belliveau et al. 2021:457). Andres Gorz introduced the concept of non-reformist reform in *Strategy for Labour* (1967). As Emilia Belliveau and her colleagues summarize, three features distinguish non-reformist reforms from system-sustaining ones:

> non-reformist reforms disrupt the capitalist status quo in ways that can benefit socialist forces; non-reformist reforms prefigure the new system 'by building popular power in the process of fighting for the reform'; reform is not the end goal in itself but it forms 'part of a larger transformative plan'.
>
> *(2021:458–9)*

Scholar-activists at the Institute for Critical Social Analysis in Berlin have developed the similar concept of 'entry projects' – 'socio-cultural learning processes of the transformation of relations and the self-transformation of the actors' (Brie 2010). In interviews I conducted there in 2012, Rainer Rilling offered the illuminating example of the struggle for free public transport:

> We tried to find fields for entry projects – for example, people should not pay for the public transit here in the town.... It's just a small suggestion. You don't have to buy something when you want to travel here in the town. You just get it. It's kind of an 'entry project,' and [when] the ruling class says 'no,' [we see] that this is an entry project because in the background there are these big questions, and that's the reason why they don't

> allow it.... And when you talk about the whole field of commons, it has to do with entry projects: the real access questions which are the beginning of self-empowerment.
>
> *(quoted in Carroll 2016:160–1)*

Obviously, free public transport, currently available in approximately 100 municipalities around the world, can be a move against the hegemony of automobility and privatized, commoditized consumption, which can contribute to a deep just transition. As with any reform, the devil is in the details, and in how the reform is articulated to other practices. In some cities, free transport is reserved for documented residents and available only in a limited area (typically the downtown core, which tends to house the affluent). Access barriers serve to reproduce class inequities, as does 'fiscal dumping', compared with funding through progressive taxation (or even elevated downtown parking fees). Judith Dellheim (2020) observes, 'a well-developed municipal *and* regional free public transport system is especially successful when it is integrated into a policy that centres on the health and wellbeing of citizens, aims to facilitate residents' active participation in society (especially from socially marginalized groups), and endeavours to increase everyone's leisure time.' She concludes that 'for a transport system based on the principles of solidarity and the protection of the environment, free public transport merely represents a single step – albeit an essential one.' What is key is that the reform opens space for greater democracy, which means that the method of reform needs to challenge 'the alienation of most people from control over their economic lives,' as Arthur McEwen has argued. 'Democratic initiatives, nonreformist reforms, cannot simply be for the people; they need to be of the people and by the people as well' (1999:18).

This example underlines the importance of bundling non-reformist reforms within a larger transformative project, so that an alternative social logic comes into play. Energy Democracy presents an example.[8] Grounded in struggles for a just energy transition in Europe (Szulecki 2018), energy democracy's three overarching goals – '*resisting* the fossil-fuel-dominant energy agenda while *reclaiming* and democratically *restructuring* energy regimes' – inform a bundle of practices that include divestment initiatives, anti-fracking protests, Indigenous activism, community solar projects, etc. (Burke and Stephens 2017, 35, 45). For Greg Albo and Lilian Yap, energy democracy means 'public ownership and control; diversity, decentralization and localization in production and control; and transparency and accountability in ecological impacts' (Albo and Yap 2016). They include in their concept democratization and participatory planning over centralized energy production and supply systems and publicly supported technological transfer to equalize renewable energy access globally. When framed in simple localist terms, energy democracy fails to connect the dots between particular communities striving for energy democracy (typically in the North) and extractive capitalism (Droubi et al. 2022), yet its uptake by many progressives, particularly through Trade Unions for Energy Democracy (TUED),[9] deepens the just-transition project.

Energy Democracy is a bundle of non-reformist reforms that can be nested within a wider war of position, to transform the correlation of political, economic and cultural forces by building an historical bloc for eco-socialist revolution. Such positional warfare creates new practices at different scales, and new solidarities, while weakening the hegemonic bloc. The components of the war of position are sometimes presented as 'transitional demands', a practice pioneered by Marx and Engels in *The Communist Manifesto*, which called, among other reforms, for 'free education for all children in public schools' (2019 [1848]). In a contemporary eco-socialist voice, Hans Baer offers this list of 'Transitional System-Challenging Reforms' that could 'facilitate a transition from the present existing capitalist world system to a democratic eco-socialist world system':

> (1) the creation of new left parties designed to capture the state; (2) emissions taxes at the sites of production; (3) public and social ownership of the means of production; (4) increasing social equality and achieving a sustainable population size; (5) workers' democracy; (6) meaningful work and shortening the work week; (7) challenging or rethinking the growth paradigm (8) energy efficiency, renewable energy sources, appropriate technology, and green jobs; (9) sustainable public transportation and travel; (10) sustainable food production and forestry; (11) resisting the culture of consumption and adopting sustainable and meaningful consumption patterns; (12) sustainable trade; and (13) sustainable settlement patterns and local communities.
>
> *(2018:13)*

Ultimately, achieving these transitional radical reforms 'would require that new left or socialist-oriented parties come to power and ensure that there is the political will that drives their implementation' (Baer and Singer 2022:88).

Creating a New Political Instrument for Eco-socialist Transformation

The prospects for eco-socialist transformation seem slim in our immediate situation. Although green-left ideas are 'scattered about, here and there … one looks in vain for a viable ecosocialist politics. As political stakes mount, prospects for revolutionary change recede' (Boggs 2021:137). To build transformational coalitions, to shift power through the conduct of a war of position – to sustain the transformation from Fossil Capitalism to eco-socialism – a political instrument capable of mobilizing power from below, challenging power on the intersecting terrains of state, economy and civil society, and building the bases for post-capitalist alternatives is indispensable. Gramsci called this instrument the Modern Prince – referring back to Niccolò Machiavelli's *The Prince*, the most significant early modern European work of strategic political theory. The projects I reviewed in Chapter 5 – GND/JT, Degrowth, Buen Vivir – each contain eco-socialist currents. The challenge lies in pulling these social forces into a coherent hegemonic project with a mass base. Such a project, as I have argued above, entails much more than militant, episodic resistance, and requires direct, ongoing engagement with capital, the state and civil society. In Gramsci's conception, the Modern Prince cannot be a concrete individual (as in Machiavelli's *Prince*), but must be 'a complex element of society in which a collective will, which has already been recognised and has to some extent asserted itself in action, begins to take concrete form' (1971:129). This 'complex element' is 'not a messianic moment deferred to the horizon, but an expansive practice within the present, as a politics that simultaneously prefigures and enacts' (Thomas 2024:4). The Modern Prince is a revolutionary party that builds, sustains and focusses that collective will on the strategic aim to found 'a new type of state' (Gramsci 1971:147, 252–253), 'a socialist state, tending to "the reabsorption of political society into civil society"' (Gramsci 1971:253; quoted in Chrysis 2024:226). In this war of position, civil society and its ideological institutions form a crucial battleground. To found a new type of state, the Modern Prince 'should intervene in the ideological institutions and ethico-political mechanisms of the concrete capitalist society and fight to alter the balance of power in favor of the proletariat and its class-allies' (Chrysis 2024:226). To prepare the conditions for self-governance, for socialist democracy, the Modern Prince must play a strong educational/cultural role as the 'bearer of a new culture' (Gramsci 1971, p. 265), 'the *collective educator* of the working masses, preparing them not only to fight against capitalism, but also to live as the citizens of a socialist democracy…' (Chrysis 2024:226).

Gramsci's strategic thinking from the 1930s has enormous relevance today. The failures of elite efforts to address the climate crisis in any serious manner tell us that capitalism will protect its conditions of existence, even as extractive accumulation degrades them. Socialist relations of production, and a greening of the forces of production equal to the task of refusing ecocide, will not arise spontaneously in the bosom of capitalism but require 'the intervention of some kind of political organization that with the support of the people conquers state power' (Harnecker 2015:174). The state, as Peter Thomas reminds us, 'remains the central antagonist of any genuinely radical politics today, and one that cannot simply be wished away' (2023:230–1). Reflecting on her many years of left activism, Marta Harnecker insisted that in the struggle for hegemony a 'new political instrument', adapted to new times, is necessary – to build solidarity among heterogeneous and fragmented progressive forces, creating both the objective capacity to act collectively and an 'internal climate that facilitates intervention into unfolding events' (2015:166). For the left, as she remarked, 'politics must be the art of making the impossible possible ... of constructing the social and political forces that are capable of changing the balance of forces to the benefit of popular movements, and making possible in the future what today appears to be impossible' (2015:167).

To accomplish this, the political instrument, the party, must be a school of popular protagonism, building people's capacities to analyze and act, and combatting the tendency for subalterns to consent spontaneously to their subordination. This means 'a focus upon the human product of revolutionary practice' (Lebowitz 2020:164), namely the 'self-changing' that occurs as people change their own circumstances (Marx 2002 [1845], thesis III). As people organize and act collectively, changing the situation, they change themselves, becoming rich in capacities to cooperate and collaborate in satisfying each other's needs, not simply to subsist but to thrive. Importantly, in building capacity for a democratic way of life, 'the revolutionary political instrument requires a horizontal relationship between the political instrument and the social movements' (Lebowitz 2020:172). The political instrument

> must be an orienting and cohering organization at the service of the social movements.... [striving] to coordinate the movements' practices into one single political project, by generating meeting spaces so that the assorted social groups can recognize each other and grow in consciousness in the specific struggles that each group has to wage in its own area: the neighborhood, university, school, factory, etc.
>
> *(Harnecker 2015:168)*

In the spaces created by the political instrument, different political strands can be braided, synthesizing the insights and vitality from Degrowth, Buen Vivir and Green New Deal initiatives with those from feminist, anti-racist and other social-justice movements. A party that builds protagonism, rather than simply contesting elections (as with liberal- and social-democratic parties) *prefigures* a democratic-socialist state. It creates a counter-hegemonic historical bloc, capable of wresting the existing state from capital, while birthing a new state through participatory-democratic institutions such as workers' and communal councils (Lebowitz 2020:166).

As I have emphasized throughout this book, the struggle for hegemony is multi-scalar. Political parties typically organize at national and sub-national levels. However, 'in a globalized world, state apparatuses, national and transnational civil societies, and institutions of the global governance regime *simultaneously* become strategic places, spaces and scales in/of hegemonic struggle' (Muhr 2024:420; Muhr 2021). Given the 'coloniality of global power', Thomas Muhr goes on to foreground the prospects for 'a counter-hegemonic historical bloc as a *Global South bloc*' – aligning progressive states and movements within just political-economic arrangements, as in the ALBA-TCP[10] alliance, articulating 'state-led internationalism' with 'movements-based transnationalism', via the ALBA Movements Political Coordination (2024:407). Although the World Social Forum (WSF), formed in 2001 as a counter-hegemonic alternative to the World Economic Forum, has not managed to develop from a discussion space to a political instrument with agentic capacity, the Progressive International, launched in 2020, has an explicit action-orientation. Its commitment to coordinated, revolutionary action reflects the advance of the 'global left' in the two decades following the emergence of the WSF. The Declaration adopted at its inaugural summit announces strong commitments to anti-capitalism, decolonization, feminism, anti-racism, Buen Vivir and internationalism as the only alternative to 'the extinction of all life in all nations across all continents' (Progressive International 2020). The Declaration asserts:

> Our aim is collective action. We are not satisfied with setting up a social network. Our activities prepare us for planetary mobilization, matching the scale of our crises to the scale of the actions that we mount against them.
>
> *(Progressive International 2020)*

As the organic crisis of global capitalism deepens, bringing greater ecological degradation and climate breakdown alongside myriad social maladies, the Progressive International, ALBA-TCP and other formations may find more space for their transformative, transnational political projects. In the meantime, binding international agreements that remove the smoke-and-mirrors approach to carbon-emissions accounting (as in the language of net zero, offsets, negative emission and carbon markets) are urgently needed. The campaign for a Fossil Fuel Non-Proliferation Treaty, a network of over 1,800 civil society organizations launched in 2019, exemplifies the possibilities.[11]

This book has focussed on Fossil Capitalism and climate crisis. Refusing ecocide is the most pressing existential issue of our time. But as the Progressive International and other examples cited throughout the book intimate, a host of movements and political currents are integral to creating a just and liveable world for all. Besides the proletariat, with its base in capitalist relations of production, the key ones, as Nancy Fraser argues, revolve around 'social reproduction,[12] the earth's ecology, political power, and ongoing infusions of wealth expropriated from racialized peoples' 2023 (2022:17). Building an historical bloc around a vast and diverse environmental proletariat means integrating 'other emancipatory currents', including feminist, anti-racist and anti-imperialist, which furnish the 'background conditions' for capitalist production itself 2023 (2022:17). A broad historical bloc portends an eco-socialism committed to gender and racial justice and decolonization, extending democratic practice throughout state, civil society and economy.

Building that bloc while pursuing tactical struggles day-to-day in the ongoing war of position is our best bet in refusing ecocide. The transformation must be 'twofold' – combining non-reformist reforms with system transformation, so that 'any initially restricted partial reforms and steps taken to overcome the capitalist system as a whole occur simultaneously' (Klein 2024:66). A capacious eco-socialist project directly confronts the trifecta of power that is at the heart of ecocide and social injustice. It offers a just, viable economic alternative to capitalism, capable of addressing the climate emergency. It provides an alternative hegemonic project capable of unifying a post-capitalist historical bloc. It challenges the geopolitical economy of ecological imperialism and opens towards a world order organized for cooperation, solidarity and peace. Our current trajectory is indeed perilous, but there is still time to correct course.

Notes

1 Notably William Morris, whose *News from Nowhere* (1890) brought a vision of a world beyond capital to a wide readership.
2 I thank Christopher Chase-Dunn for this apt term.
3 Saitō's important work has been critiqued by Huber and Phillips, who suggest that the analysis misinterprets Marx's notes on the Russian *mir* (collectively organized peasant communities). In those notes, Marx did not endorse Degrowth. He argued that the *mir* 'could leapfrog capitalist development because capitalist development had occurred elsewhere, in the same way that many poor countries have jumped directly to adoption of mobile phones without having to pass through the stages of telegraphy or landlines' (Huber and Phillips 2024). David Schwartzman (2023) observes that Saitō ignores the left critique of Degrowth (which I have presented above). And, disappointingly, 'does not systematically deconstruct the degrowth discourse' employing the normative distinction between good and bad growth 'in the context of a strategy to reach the goal of degrowth communism.' Notwithstanding these critiques, I find great insight in Saitō's analysis.
4 The distinction between 'effectiveness' and 'efficiency' is relevant here. Climate Capitalism's hegemonic narrative emphasizes the latter, which gives license to continuing emissions as long as there are 'efficiency gains'. Instead, 'effectiveness must take precedence over efficiency in reducing emissions. That means abandoning the fetish of the price mechanism in order to plan how the remaining dirty resources will be used in the service of clean infrastructure. Such planning must have international reach, since the greatest opportunities for energy-supply decarbonation are located in the Global South' (Durand 2021).
5 'Contemporary economic planning can draw on techniques from meteorology' and other advances in the forces of production, including those stemming from climate modelling. 'The data density of the contemporary world, paired with the algorithms climate scientists have designed to handle it, greatly expands planning capacities' (Vettese and Pendergrass 2022:129).
6 The process of bringing decision-making to the level of communities, rather than centralizing power in distant sites, known as subsidiarity, informs Walden Bello's conception of deglobalization. He notes that subsidiarity encourages 'production of goods at the level of the community and at the national level if this can be done at reasonable cost in order to preserve community' (Bello 2005:114).

7 See the documentary, 'This is What Democracy Looks Like' (Friedberg and Rowley 2000), available at https://www.youtube.com/watch?v=_-LfN3V3id8. Hayden's speech begins at 64 minutes.
8 See also Goodman and Morton's (2023) discussion of climate democracy.
9 See the TUED website https://www.tuedglobal.org/, and see Sweeney (2023) for an example of TUED policy research, in this paper calling for a *left* Global Green New Deal that features public ownership of the power sector.
10 This Castilian acronym translates into English as the Bolivarian Alliance for the Peoples of Our America – People's Trade Agreement (Muhr 2024:407). ALBA includes over 400 organizations and popular movements from 25 countries (Muhr 2024:406), in a cooperative regional framework that has, at its heart, solidaristic fair trade, in sharp contrast to neoliberal 'free trade.'
11 See https://fossilfueltreaty.org/.
12 By social reproduction, Fraser means 'the forms of provisioning, caregiving, and interaction that produce and sustain human beings and social bonds' (2022:9).

References

Albo, Greg. 2006. "The Limits of Eco-Localism: Scale, Strategy, Socialism." *Socialist Register* 43:337–63.

Albo, Greg, and Lilian Yap. 2016. "From the Tar Sands to 'Green Jobs'? Work and Ecological Justice." *The Bullet*. Retrieved April 4, 2024 (https://socialistproject.ca/2016/07/b1280/).

Azmanova, Albena. 2019. "The Paradox of Emancipation: Populism, Democracy and the Soul of the Left." *Philosophy & Social Criticism* 45(9–10):1186–1207. doi: 10.1177/0191453719872291

Baer, Hans. 2018. *Democratic Eco-Socialism as a Real Utopia: Transitioning to an Alternative World System*. New York: Berghahn.

Baer, Hans, and Merrill Singer. 2022. "Can Ecological Modernisation Contain Climate Change?" *The Journal of Australian Political Economy* 90:75–91.

Bakx, Kyle. 2024. "$34B Trans Mountain Expansion Pipeline Begins Filling with Oil with First Shipments before Canada Day." *CBC News*. Retrieved April 2, 2024 (https://www.cbc.ca/news/canada/calgary/trans-mountain-expansion-begins-1.7150343).

Belliveau, Emilia, James K. Rowe, and Jessica Dempsey. 2021. "Fossil Fuel Divestment, Non-Reformist Reforms, and Anti-Capitalist Strategy." Pp. 453–78 in *Regime of Obstruction*, edited by W. K. Carroll. Edmonton: AU Press.

Bello, Walden. 2005. *Deglobalization: Ideas for a New World Economy*. New Updated. London: Zed Books.

Boggs, Carl. 2021. *Fugitive Politics: The Struggle for Ecological Sanity*. 1st ed. New York: Routledge.

Brie, Michael. 2010. *Entry Projects to a Politics of Solidarity*. Berlin: Rosa Luxemburg Foundation.

Brownhill, Leigh, Salvatore Engel-DI Mauro, Terran Giacommi, Ana Isla, Michael Löwy, and Terisa E. Turner. 2022. "Preface: An Introduction to Ecosocialism." Pp. 1–5 in *The Routledge Handbook on Ecosocialism*, edited by Leigh Brownhill, Salvatore Engel-Di Mauro and Terran Giacomini. London: Routledge.

Buck, Holly Jean. 2019. *After Geoengineering*. London: Verso.

Burke, Matthew J., and Jennie C. Stephens. 2017. "Energy Democracy: Goals and Policy Instruments for Sociotechnical Transitions." *Energy Research & Social Science* 33:35–48. doi: 10.1016/j.erss.2017.09.024

Burkett, Paul. 2020. "An Eco-Revolutionary Tipping Point?" *Revista Internacional De Salarios Dignos* 2(1):82–94.

Candeias, Mario. 2013. *Green Transformation: Competing Strategic Projects*. Berlin: Rosa Luxemburg Foundation. Accessed August 24, 2024 (https://rosalux.nyc/green-transformation/).

Carroll, William K. 2010. *The Making of a Transnational Capitalist Class*. London: Zed Books.

Carroll, William K. 2015. "Robust Radicalism." *Review of Radical Political Economics* 47(4):663–68. doi: 10.1177/0486613415574272

Carroll, William K. 2016. *Expose, Oppose, Propose: Alternative Policy Groups and the Struggle for Global Justice*. London and Halifax: Zed Books and Fernwood Publishing.

Carroll, William K. 2022. "Marx Matters, in Theory and Practice: Reflections from the Corporate Mapping Project." Pp. 198–225 in *Marx Matters*, edited by D. Fasenfest. Leiden: Brill.

Carroll, William K., Nicolas Graham, Michael Lang, Kevin McCartney, and Zoe Yunker. 2021. "Fossil Capital's Reach into Civil Society: The Architecture of Climate Change Denialism." Pp. 171–96 in *Regime of Obstruction: How Corporate Power Blocks Energy Democracy*, edited by William K. Carroll. Edmonton: AU Press.

Carroll, William K., Jouke Huijzer, and J. P. Sapinski. 2023. "Corporate Networks." Pp. 336–49 in *The Sage Handbook of Social Network Analysis*, edited by J. McLevey, J. Scott, and P. Carrington. London: Sage.

CBC News. 2023. "Despite Opposition and Environmental Violations, Major B.C. Pipeline Project Nearly Complete." *CBC News*. Retrieved May 27, 2024 (https://www.cbc.ca/news/canada/british-columbia/coastal-gaslink-fully-installed-1.7013274).

Chambers, Collin L. 2021. "A Critique of the 'Socio-ecological Fix' and towards Revolutionary Rupture." *Area* 53(1):114–21. doi: 10.1111/area.12668

Chen, Ying. 2023. "Degrowth: What's in a Name?" *Monthly Review* 75(3):160–70.

Chrysis, Alexandros. 2024. "The Modern Prince and Revolutionary Strategy." Pp. 219–37 in *The Elgar Companion to Antonio Gramsci*, edited by W. K. Carroll. Cheltenham, UK: Edward Elgar.

Ciplet, David. 2022. "Transition Coalitions: Toward a Theory of Transformative Just Transitions." *Environmental Sociology* 8(3):315–30. doi: 10.1080/23251042.2022.2031512

Ciplet, David. 2023. *Beyond the Mitigation-Adaptation Binary: Advancing Global Climate Justice through the Six Pillars of Climate Governance*. Philadelphia: American Sociological Association. Retrieved June 2, 2024 (https://ssrn.com/abstract=4534319).

Coulthard, Glen. 2013. "For Our Nations to Live, Capitalism Must Die." *Unsettling America*. Accessed August 28, 2024 (https://unsettlingamerica.wordpress.com/2013/11/05/for-our-nations-to-live-capitalism-must-die/)

De Smet, Brecht. 2014. "Alterglobalization and the Limits of Prefigurative Politics." Pp. 311–16 in *Collaborative Projects*, edited by A. Blunden. Leiden: Brill.

Dellheim, Judith. 2020. "Public Transport at Zero Cost." *Rosa Luxemburg Stiftung*. Retrieved April 3, 2024 (https://www.rosalux.de/en/news/id/42870/public-transport-at-zero-cost).

Droubi, Sufyan, Raphael J. Heffron, and Darren McCauley. 2022. "A Critical Review of Energy Democracy: A Failure to Deliver Justice?" *Energy Research & Social Science* 86:102444. doi: 10.1016/j.erss.2021.102444

Dunbar-Ortiz, Roxanne Amanda. 2016. "The Relationship between Marxism and Indigenous Struggles and Implications of the Theoretical Framework for International Indigenous Struggles." *Historical Materialism* 24(3):76–91. doi: 10.1163/1569206X-12341485

Durand, Cedric. 2021. "Energy Dilemma." *New Left Review – Sidecar*. Retrieved February 29, 2024 (https://newleftreview.org/sidecar/posts/energy-dilemma).

Eaton, Emily. 2021. "Approaches to Energy Transitions: Carbon Pricing, Managed Decline, and/or Green New Deal?" *Geography Compass* 15(2):e12554. doi: 10.1111/gec3.12554

Foster, John Bellamy. 2010. "Why Ecological Revolution?" *Monthly Review* 61(8):1–18. doi: 10.14452/MR-061-08-2010-01_1

Foster, John Bellamy, and Jia Keqing. 2023. "Ecological Marxism." *Monthly Review* 75(4):35–50. doi: 10.14452/MR-075-04-2023-08_4

Foster, John Bellamy, and Batuhan Sarican. 2023. "'Monthly Review' and the Environment." *Monthly Review* 75(6):1–17.

Fraser, Nancy. 2022. *Cannibal Capitalism: How Our System Is Devouring Democracy, Care, and the Planet – and What We Can Do about It*. London; New York: Verso.

Friedberg, Jill, and Rick Rowley. 2000. *This Is What Democracy Looks Like*. Seattle: Corrugated Films and Big Noise Films. Acccessed August 28, 2024 (http://www.corrugate.org/this-is-what-democracy-looks-like.html)

Gindin, Sam. 2019. "Socialism for Realists." *Catalyst* 2(3):7–45.

Goodman, James, and Tom Morton. 2023. "'Rage into Action': From Carbon Democracy to Climate Democracy?" *Globalizations* 20(8):1257–76. doi: 10.1080/14747731.2023.2270193

Gorz, André. 1967. *Strategy for Labor: A Radical Proposal*. Boston: Beacon Press.

Graham, Nicolas. 2023. "Planning and the Ecosocialist Mode of Cooperation." *Monthly Review* 75(3):126–41. doi: 10.14452/MR-075-03-2023-07_9

Gramsci, Antonio. 1971. *Selections from the Prison Notebooks*. edited by Q. Hoare and G. Nowell-Smith. New York: Lawrence & Wishart.

Habermas, Jurgen. 1987. *The Theory of Communicative Action*. Vol. 2. Boston: Beacon Press.

Haines, Christian P. 2023. "Planetary Utopianism: Geoengineering, Speculative Fiction, and the Planetary Turn." *Textual Practice* 37(9):1343–63. doi: 10.1080/0950236X.2023.2213062

Hall, Stuart. 1988. *The Hard Road to Renewal*. London: Verso.

Harnecker, Marta. 2015. *A World to Build: New Paths toward Twenty-First Century Socialism*. New York: Monthly Review Press.

Hickel, Jason. 2023. "On Technology and Degrowth." *Monthly Review* 75(3):44–50. doi: 10.14452/MR-075-03-2023-07_3

Huber, Matthew T. 2022. *Climate Change as Class War*. London: Verso.

Huber, Matthew T., and Leigh Phillips. 2024. "Kohei Saitō's 'Start From Scratch' Degrowth Communism." *Jacobin*. Retrieved March 25, 2024 (https://jacobin.com/2024/03/kohei-saito-degrowth-communism-environment-marxism).

Işıkara, Güney, and Özgür Narin. 2023. "Degrowth and Socialism: Notes on Some Critical Junctures." *Monthly Review* 75(3):30–43. doi: 10.14452/MR-075-03-2023-07_2

Kallis, Giorgos. 2017. "Radical Dematerialization and Degrowth." *Philosophical Transactions of the Royal Society A: Mathematical, Physical and Engineering Sciences* 375(2095):20160383. doi: 10.1098/rsta.2016.0383

Klein, Dieter. 2024. *Within and Beyond Capitalism: A Twofold Transformation*. London: Routledge.

Klein, Naomi. 2007. *The Shock Doctrine*. Toronto: Alfred A. Knopf Canada.

Klein, Naomi. 2014. *This Changes Everything: Capitalism vs. the Climate*. Toronto: Alfred A. Knopf Canada.

Lebowitz, Michael A. 1995. *Beyond Capital: Marx's Political Economy of the Working Class*. London: Macmillan.

Lebowitz, Michael A. 2020. *Between Capitalism and Community*. New York: Monthly Review Press.

Löwy, Michael. 2023. "Nine Theses on Ecosocialist Degrowth." *Monthly Review* 75(3):154–59.

Magdoff, Fred, and John Bellamy Foster. 2011. *What Every Environmentalist Needs to Know about Capitalism*. New York: Monthly Review Press.

Malm, Andreas, and Wim Carton. 2021. "Seize the Means of Carbon Removal: The Political Economy of Direct Air Capture." *Historical Materialism* 29(1):3–48. doi: 10.1163/1569206X-29012021

Marx, Karl. 2022. *Critique of the Gotha Program*. Oakland, CA: PM Press.

Marx, Karl. 1976. *Capital*. Vol. 1. London: Penguin.

Marx, Karl. 2002. "Theses on Feuerbach." *Marx/Engels Internet Archive*. Retrieved June 2, 2024 (https://www.marxists.org/archive/marx/works/1845/theses/theses.htm).

Marx, Karl. 2009. *The Poverty of Philosophy. Marx/Engels Internet Archive*. Retrieved August 28, 2024 (https://www.marxists.org/archive/marx/works/download/pdf/Poverty-Philosophy.pdf).

Marx, Karl, and Friedrich Engels. 2019. "Manifesto of the Communist Party." *Marx/Engels Internet Archive*. Retrieved May 14, 2023 (https://www.marxists.org/archive/marx/works/1848/communist-manifesto/).

Maslin, Mark A., John Lang, and Fiona Harvey. 2023. "A Short History of the Successes and Failures of the International Climate Change Negotiations." *UCL Open Environment* 5. doi: 10.14324/111.444/ucloe.000059

Mau, Soren. 2023. *Mute Compulsion: A Marxist Theory of the Economic Power of Capital*. London: Verso.

McEwen, Arthur. 1999. *Neo-Liberalism or Democracy? Economic Strategy, Markets, and Alternatives*. London: Zed Books.

Morris, William. 1890. *News from Nowhere*. London: Commonweal. Accessed August 28, 2024 (https://www.marxists.org/archive/morris/works/1890/nowhere/index.htm).

Muhr, Thomas. 2021. "Counter-Hegemonic Strategy from the Global South: A Pluri-Scalar War of Position." *Socialism and Democracy* 35(2–3):214–40. doi: 10.1080/08854300.2021.1994295

Muhr, Thomas. 2024. "Gramsci and Hegemonic Struggle in a Globalized World." Pp. 406–26 in *The Elgar Companion to Antonio Gramsci*, edited by W. K. Carroll. Cheltenham, UK; Northampton, MA: Edward Elgar Publishing.

O'Neill, Daniel W., Andrew L. Fanning, William F. Lamb, and Julia K. Steinberger. 2018. "A Good Life for All within Planetary Boundaries." *Nature Sustainability* 1(2):88–95. doi: 10.1038/s41893-018-0021-4

Parenti, Christian. 2022. "A Left Defense of Carbon Dioxide Removal: The State Must Be Forced to Deploy Civilization-Saving Technology." Pp. 130–42 in *Has it Come to This? The Promises and Perils of Geoengineering on the Brink*, edited by J. P. Sapinski, H. J. Buck, and A. Malm. New Brunswick, NJ: Rutgers University Press.

Progressive International. 2020. "Declaration." *Progressive International.* Retrieved April 8, 2024 (https://progressive.international/declaration/en).

Saitō, Kōhei. 2022. *Marx in the Anthropocene: Towards the Idea of Degrowth Communism.* Cambridge; New York: Cambridge University Press.

Satgar, Vishwas. 2018. "The Anthropocene and Imperial Ecocide: Prospects for Just Transitions." Pp. 47–67 in *The Climate Crisis*, Edited by Vishwas Satgar. Johannesburg: Wits University Press.

Schmelzer, Matthias, and Elena Hofferberth. 2023. "Democratic Planning for Degrowth." *Monthly Review* 75(3):142–53. doi: 10.14452/MR-075-03-2023-07_10

Schoppek, Dorothea Elena. 2020. "How Far Is Degrowth a Really Revolutionary Counter Movement to Neoliberalism?" *Environmental Values* 29(2):131–51. doi: 10.3197/096327119X15579936382491

Schwartzman, David. 2023. "Critical Comments on Kohei Saitō's View of 'Degrowth Communism.'" *Climate & Capitalism.* Retrieved March 25, 2024 (https://climateandcapitalism.com/2023/06/25/critical-comments-on-kohei-saitos-view-of-degrowth-communism/).

Sharzer, Greg. 2012. *No Local: Why Small-Scale Alternatives Won't Change the World.* Winchester, UK: Zero Books.

Smucker, Jonathan. 2017. *Hegemony How-To: A Roadmap for Radicals.* Chico, CA: AK Press.

Sweeney, Sean. 2023. *Beyond Recovery: The Global Green New Deal and Public Ownership of Energy.* 16. New York: Rosa Luxemburg Stiftung.

Swift, Richard. 2016. *S.O.S. Alternatives to Capitalism.* Second Edition. Toronto: Between the Lines.

Szulecki, Kacper. 2018. "Conceptualizing Energy Democracy." *Environmental Politics* 27(1):21–41. doi: 10.1080/09644016.2017.1387294

Thomas, Peter D. 2023. *Radical Politics.* New York: Oxford University Press.

Thomas, Peter D. 2024. "New Orders: Hegemony as a Method of Political Work." Presented at Practices of Resistance in Times of Crisis, 19 April. Toronto: International Gramsci Society.

Vettese, Troy, and Drew Pendergrass. 2022. *Half-Earth Socialism.* London: Verso.

INDEX

Pages in *italics* refer to figures, pages in **bold** refer to tables, and pages followed by "n" refer to notes.